Synthesis Lectures on Ocean Systems Engineering

Series Editor

Nikolas Xiros, University of New Orleans, New Orleans, LA, USA

The series publishes short books on state-of-the-art research and applications in related and interdependent areas of design, construction, maintenance and operation of marine vessels and structures as well as ocean and oceanic engineering.

Alexander Arnfinn Olsen

Offshore Access Gangways

Alexander Arnfinn Olsen ⓘ
Southampton, UK

ISSN 2692-4420　　　　　　ISSN 2692-4471　(electronic)
Synthesis Lectures on Ocean Systems Engineering
ISBN 978-3-031-74782-3　　　ISBN 978-3-031-74783-0　(eBook)
https://doi.org/10.1007/978-3-031-74783-0

© The Editor(s) (if applicable) and The Author(s), under exclusive license to Springer Nature Switzerland AG 2025

This work is subject to copyright. All rights are solely and exclusively licensed by the Publisher, whether the whole or part of the material is concerned, specifically the rights of translation, reprinting, reuse of illustrations, recitation, broadcasting, reproduction on microfilms or in any other physical way, and transmission or information storage and retrieval, electronic adaptation, computer software, or by similar or dissimilar methodology now known or hereafter developed.
The use of general descriptive names, registered names, trademarks, service marks, etc. in this publication does not imply, even in the absence of a specific statement, that such names are exempt from the relevant protective laws and regulations and therefore free for general use.
The publisher, the authors and the editors are safe to assume that the advice and information in this book are believed to be true and accurate at the date of publication. Neither the publisher nor the authors or the editors give a warranty, expressed or implied, with respect to the material contained herein or for any errors or omissions that may have been made. The publisher remains neutral with regard to jurisdictional claims in published maps and institutional affiliations.

This Springer imprint is published by the registered company Springer Nature Switzerland AG
The registered company address is: Gewerbestrasse 11, 6330 Cham, Switzerland

If disposing of this product, please recycle the paper.

Preface

As the demand for a safe and efficient walk to work (W2W) approach prevails in the offshore oil and gas industry, it is envisaged that there will be a significant growth in the use of offshore access gangway systems for the manning and transfer of personnel to and from offshore facilities.

Noting the special design and operational characteristics of offshore access gangway systems, this short reference guide has been developed based on existing Class guidance to provide provisions for the certification of offshore access gangway systems used for connecting two offshore units to transfer offshore personnel on a temporary basis. The requirements contained within this reference guide include either via direct inclusion or a reference to Class Rules or guides as well as relevant recognised international Regulations.

This reference guide pertains to the relevant international statutory Regulations and guidelines that are considered applicable and therefore worthy of note.

> While it is the intent of this reference guide to be consistent with these Regulations and guidelines, it is the ultimate responsibility of the reader of this reference guide to refer to the most recent text of those Regulations and guidelines. Subsequently, this reference guide should be read in conjunction with other Rules published by Class and the recognised international Regulations.

This reference guide will serve as a useful reference for the designers, builders, owners, and operators of offshore access gangway systems.

Southampton, UK   Alexander Arnfinn Olsen

Contents

1 **Scope and Conditions of Certification** 1
   1.1   Application 1
   1.2   Scope 1
   1.3   Class Notations 2
   1.4   Existing Offshore Access Gangway Systems 3
   1.5   Class Approvals 3
   1.6   Other Regulations 3
       1.6.1   International and Other Regulations 3
       1.6.2   Governmental Regulations 4
   1.7   Submissions of Plans 4
   1.8   Units 4

2 **General Provisions** 5
   2.1   General 5
       2.1.1   Offshore Access Gangways 5
       2.1.2   Category of Gangways 7
       2.1.3   Types of Motion Compensation Systems 7
   2.2   Submissions of Plans and Design Data 8
       2.2.1   General 8
       2.2.2   Information to Be Submitted to Class 8
   2.3   Loading, Handling and Securing 11
   2.4   Terms and Definitions 11
   2.5   Certifications of Components 13

3 **Structural Requirements** 17
   3.1   General 17
   3.2   Materials 17
       3.2.1   Steel 17
       3.2.2   Aluminium Alloys 18
       3.2.3   Effective Corrosion Control 18

3.3	Design Loads and Conditions		19
	3.3.1	Design Loads	19
	3.3.2	Design Conditions	23
3.4	Strength Assessment		26
	3.4.1	General	26
	3.4.2	Allowable Stress Assessment Criteria for ULS and ALS	27
	3.4.3	Fatigue Assessment Criteria for FLS	29
	3.4.4	Serviceability Limit State (SLS)	30
3.5	Landing Mechanism		32
3.6	Pedestals, Foundations, and Supporting Structures		32
3.7	Slewing Mechanism		33
3.8	Telescoping Mechanism		33
3.9	Wire Rope		33
3.10	Loose Gears		33

4 Machinery and Systems 35
4.1	General		35
4.2	Materials		35
4.3	Electrical Systems		35
4.4	Piping Systems		35
	4.4.1	Hydraulic Systems	36
4.5	Pressure Vessels		36
4.6	Rotating Machines		36
4.7	Computer-Based Control Systems		37
4.8	Motion Compensation Systems		37
4.9	Low Temperature Operation		38
4.10	Hazardous Locations		38
4.11	Blocking Mechanism		38
4.12	Electronics and Communications		39
4.13	Emergency Recovery		39
4.14	Safety Systems and Arrangements		39
	4.14.1	General	39
	4.14.2	Monitoring Systems	40
	4.14.3	Alarm System	40
	4.14.4	Handrails and Grids	40
	4.14.5	Slip Resistant Surface	41
	4.14.6	Landing Area	41
	4.14.7	Break-Away System	41
	4.14.8	Fire Protection	41
	4.14.9	Control Cabin Protection	41

	4.15	Visual Aids	42
		4.15.1 Markings	42
		4.15.2 Lighting System	42
5	**Testing and Surveys**		**43**
	5.1	General	43
	5.2	Surveys During Construction	43
		5.2.1 General	43
		5.2.2 Functionality Testing	44
	5.3	Load Testing	44
		5.3.1 Load Testing Procedures	44
		5.3.2 Test Loads	46
	5.4	Surveys During Installation	47
	5.5	Surveys After Construction	48
		5.5.1 Annual Survey	48
		5.5.2 Retesting Survey	49
		5.5.3 Repairs and Alterations	50
		5.5.4 Reinstallation Survey	51
	5.6	Slewing Ring Surveys	51
	5.7	Inspection of Wire Rope	51
	5.8	Monthly Inspection by Vessel's Personnel	52
6	**Risk Assessment and Register for Offshore Access Gangways**		**53**
	6.1	General	53
	6.2	Register for Offshore Access Gangways	54
	6.3	Certificates and Forms	54
	6.4	Asset Owner's Overhaul and Inspection Record	55
	6.5	Repairs and Alterations	55
	6.6	Additions of New Gear and Wire Rope	55
Appendix			**57**
References			**71**

Abbreviations and Acronyms

AISC	American Institute of Steel Construction
ALS	Accidental Limit States
ASD	Allowable Stress Design
DPS	Dynamic Positioning System
EER	Escape, Evacuation, and Rescue
FLS	Fatigue Limit State
FMEA	Failure Modes and Effect Analysis
FMECA	Failure Mode Effect and Criticality Analysis
HAZID	Hazard Identification
HAZOP	Hazard and Operability
IMO	International Maritime Organisation
ISO	International Standards Organisation
MKS	Metre-Kilogram-Second
MODU	Mobile Offshore and Drilling Unit
MOU	Mobile Offshore Unit
MSC	Maritime Safety Committee
NDT	Non-destructive Testing
SI	International System of Units
SLS	Serviceability Limit State
SOLAS	International Convention for the Safety of Life at Sea
SWL	Safe Working Load
UK HSE	UK Health and Safety Executive
ULS	Ultimate Limit States
W2W	Walk to Work

List of Figures

Fig. 2.1	Example configuration of an offshore access gangway	6
Fig. 3.1	Type I gangway: distributed live loads	19
Fig. 3.2	Type I gangway: live loads for emergency lift-off condition	20
Fig. 3.3	Type II gangway: live load for gangway supported at two ends	20
Fig. 3.4	Type II gangway: live load for uplift or cantilever condition	21
Fig. 3.5	Type II gangway: live load for emergency lift-off condition	21
Fig. 3.6	Relative deflection	31

List of Tables

Table 2.1	Examples of primary structure	13
Table 2.2	Examples of critical machinery components	13
Table 2.3	Offshore access gangway components certification[a]	14
Table 3.1	Design loads applied to the design conditions	24
Table 3.2	Structural strength assessment	27
Table 3.3	Allowable stress coefficients, S_c	28
Table 3.4	Fatigue design factors	30

Scope and Conditions of Certification

1.1 Application

This reference guide contains provisions for the certification of offshore access gangway systems installed aboard vessels, offshore floating units, and fixed offshore units. If specifically requested by the asset owner, and agreed by the Flag State Administration, the information contained within this guide may also be used as a basis for applying for acceptance or certification under the requirements of the Flag State Administration. Asset owners who desire to have offshore access gangway systems evaluated for compliance with National Regulations should contact the relevant classification society responsible for classing the said asset.

1.2 Scope

This reference guide outlines the requirements for certification of all offshore access gangway systems installed on a vessel or offshore floating unit, including, but not limited to the following routine gangway operations from:

- Vessel to vessel,
- Vessel to offshore floating unit,
- Offshore floating unit to vessel,
- Offshore floating unit to offshore floating unit,
- Vessel to fixed offshore unit, and
- Offshore floating unit to fixed offshore unit.

This reference guide also applies to connected vessels or offshore floating units that maintain position by means of Dynamic Positioning System (DPS), mooring system, or any sophisticated station keeping system, and offshore fixed units.

In accordance with the established rules pertaining to the design and construction of maritime gangways, these are to be designed and certified taking into consideration the vessel/unit where they will be installed. When the gangway is reinstalled to a new vessel/unit, either within the same field or in a different operating area, the strength of the gangway must be reassessed to satisfy that the gangway will remain in compliance with applicable requirements. The design reviews and surveys related to the reinstallation are to be completed in accordance with Flag State Administration and Class Rules.

1.3 Class Notations

The Class notation 'Certification of offshore access gangways,' or equivalent, is issued under the oversight of Class and follows the typical register procedure in the Appendix. A vessel or unit which has a recognised Register of offshore access gangway systems permanently installed, may be distinguished by an additional class notation. For example, the American Bureau of Shipping (ABS) provides the following notation **GRC (Type I or II, PS, or AS)**, which may be subcategorised accordingly:

Type I signifies that the gangway system permits unrestricted flow of personnel transfer within the capacity limitation and is supported at both ends.

Type II signifies that the gangway system permits limited flow of personnel transfer.

PS signifies that the vessel or unit has an installed passive motion compensation gangway system designed, constructed, and tested in accordance with the respective requirements established by ABS. Refer to specific Class requirements for details.

AS signifies that the vessel or unit has an installed active or full active motion compensation gangway system designed, constructed, and tested in accordance with the respective requirements established by ABS. Refer to specific Class requirements for details.

To use this example, a vessel with an active motion compensation gangway system and the gangway allows personnel to transfer freely is to be assigned the notation **GRC (Type I-AS)**.

In the case of a vessel with more than one gangway, separate reviews and surveys for each gangway will typically be required, whereas multiple gangway systems may be included in the same notation.

1.4 Existing Offshore Access Gangway Systems

For an existing offshore access gangway system, submission of information as required in the respective chapters and sections of this guide, with verification of material, is required. An existing gangway system may be certified subject to a satisfactory plan review, condition survey, operational tests (including luffing, slewing, telescoping), test of safety devices, and proof testing of the offshore access gangway systems as outlined in this guide. The condition survey is to include inspection for excessive wear, damage, corrosion, and fractures. Non-destructive testing or verification of materials may be required at the discretion of the surveyor. All mechanical, electrical, and piping systems and components are to be examined as deemed necessary by the attending surveyor.

1.5 Class Approvals

Upon application by the manufacturer, each model of a type of offshore access gangway is to be design assessed as described in this guide. For this purpose, each design of an offshore access gangway type is to be approved as outlined with the requirements of this guide. The type testing specified in Chap. 5, Sect. 5.3, is to be conducted in accordance with an approved test schedule and is to be witnessed by a surveyor. Offshore access gangways so approved may be applied to Class for listing on the respective website for 'products design assessed' or equivalent. Once listed, and subject to renewal and updating of the classification certificate, offshore access gangway particulars are usually not required to be submitted to Class each time the gangway is proposed for use on board a vessel.

1.6 Other Regulations

1.6.1 International and Other Regulations

While this reference guide covers the requirements for the certification of offshore access gangway system and their equipment, the attention of asset owners, designers, and builders is directed to the regulations of international, governmental, and other authorities dealing with those requirements in addition to or over and above the classification requirements. Where authorised by the Flag State Administration of a country signatory thereto and upon request of the asset owners of a certified offshore access gangway system or one intended to be certified, upon request Class may survey for compliance with the provision of International and Governmental conventions and codes, as applicable.

1.6.2 Governmental Regulations

Where authorised by a government agency and upon request of the asset owner of a new or existing offshore access gangway system, Class will typically survey and certify a classed offshore access gangway system, or one intended to be classed for compliance with specific regulations of that government on their behalf.

1.7 Submissions of Plans

Each chapter of this reference guide identifies a list of offshore access gangway system components that are required for the certification of the offshore access gangway system. In most cases, manufacturers' components and system related drawings, calculations, and documentations are required to be submitted to substantiate the design of the system or component. In these cases, upon satisfactory completion of Class review of the manufacturers' submittals, Class engineers will issue a review letter. This letter, in conjunction with the submitted package may be used, and referenced during surveys and subsequently issued reports by attending Class surveyors. Upon satisfactory completion of all required engineering and survey processes, Class may issue the Certificate to the offshore access gangway system.

1.8 Units

This reference guide is written using three systems of units: SI units, MKS units and US customary units. Each system is to be used independently of any other system. Unless indicated otherwise, the format of presentation of the three systems of units is as follows:
 SI units (MKS units, US customary units).

General Provisions

2.1 General

This reference guide outlines the requirements for the certification of offshore access gangway systems installed aboard vessels or floating/fixed offshore units. The guide covers the offshore access gangway, its foundation, and supporting structures (e.g., tower, pedestal, etc.). However, the supporting structure under the deck that is integrated into the vessel/unit hull is not within the scope of this reference. It should be noted that any Class approval and certification of the offshore access gangway system will be limited to the reviewed conditions.

The information and guidance provided herein is applicable to offshore access gangways used for connecting two offshore units or vessels to transfer offshore personnel on a temporary basis. A gangway intended to operate at an inclination greater than 30 degrees is not within scope. A gangway operated as a primary means of escape or installed where potential hazardous utility functions are attached is subject to special consideration in consultation with the appropriate Flag State Administration.

2.1.1 Offshore Access Gangways

In general, an offshore access gangway consists of various parts, which are primarily dependent on the level of motion compensation. Figure 2.1 depicts one type of offshore access gangway that includes a non-extended part connected to the supporting structure and a telescopic part with a landing and connecting mechanism (e.g., landing device/cone).

The supporting structure of the gangway may consist of a frame with heavy steel members, a base structure typically composed of thick welded plates, and a slewing ring.

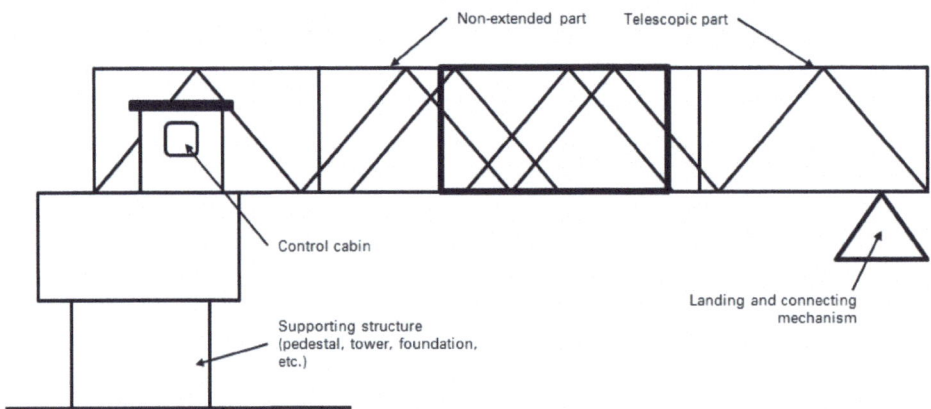

Fig. 2.1 Example configuration of an offshore access gangway

Typically, the supporting structure is integrated into the vessel deck of the vessel/unit. An offshore access gangway includes associated mechanical, hydraulic, electrical, and safety systems as well as one or more control cabin/station.

Typically, an offshore access gangway rests in a parked position on a support "cradle" on the deck of the vessel/unit. When required, the gangway is lifted, slewed, and extended to the required landing position, typically using hydraulic mechanisms. In some cases, a motion compensation system is used to reduce the lift-off motions during connection or while connected. Once the landing cone arrives at the desired position on the adjacent offshore unit, it will be fastened to that unit. In some cases, the gangway landing end may be pushed in between horizontal constraints in the landing area without fastening or operated as a cantilever, the gangway normally provides a mechanism to keep a "form closed" connection. When connected, the telescopic part is free to slide within the non-extended part, with limits set for the maximum and minimum sliding positions.

If the operating limits are exceeded, an alarm is to sound and the gangway will be lifted manually during normal operation, or lifted automatically for emergency lift, and then retrieved and returned to the support cradle.

2.1.2 Category of Gangways

An offshore access gangway is to be categorised as one of the two types in the sections below.

2.1.2.1 Type I: Unrestricted Flow of Personnel Transfer

(1) This type of gangway is to be used for unrestricted flow of personnel transfer between the connected units (i.e., an unlimited number of personnel may pass the gangway at any given time so long as the total load is within the gangway's design load capacity).
(2) The gangway is supported at both ends in all three axis directions.
(3) The telescopic end of the gangway is to be capable of being readily retracted and moving to its stowed position safely within a short period of time when, for example, design conditions are expected to be exceeded.

2.1.2.2 Type II: Limited Flow of Personnel Transfer

(1) This type of gangway is to be used for limited flow of personnel transfer between the connected units through manual or automatic flow control (i.e., only a specified number of personnel may be present on the gangway at any given time, based on the gangway design).
(2) The gangway is supported at least at one end in all three axis directions.
(3) The gangways are to be provided with the self-detach devices for the telescopic end and are to be capable of moving the gangway to its stowed position safely within a short period of time when, for example, design conditions are expected to be exceeded.

2.1.3 Types of Motion Compensation Systems

The following types of motion compensation systems are considered in this reference guide.

2.1.3.1 Full Active Motion Compensation System

A system that all drive mechanisms present (slewing, luffing, telescoping) are actively controlled and compensates for all degrees of vessel motions on the gangway structure.

2.1.3.2 Active Motion Compensation System

A system that uses motion sensor signals in the control system and external energy to compensate for the effects of the vessel (or unit) motions on the offshore access gangway.

2.1.3.3 Passive Motion Compensation System

A system with integrated features that allow the gangway to maintain the relative motions between vessels (or units) does not make use of motion sensor signals or external systems in the control system.

2.2 Submissions of Plans and Design Data

2.2.1 General

Plans showing the arrangements and details of the offshore access gangway system are to be submitted for review before fabrication begins. These plans are to clearly indicate the scantlings, materials, joint details, and welding. Plans should generally be submitted electronically to Class; however, hard copies will also be accepted.

2.2.2 Information to Be Submitted to Class

The following plans and supporting data are to be submitted for review and approval as applicable.

2.2.2.1 Offshore Access Gangway Structures

(1) General arrangement, assembly plans, and description of operating procedures and design service temperature.
(2) Applicable design loads (i.e., dead, live, and dynamic loads) including details for gangway stiffness; environmental loads including the effects of wind, snow, and ice; loads caused by luffing, slewing, and telescoping operations; loads due to static list and/or trim of the vessel or unit; loads due to vessel or unit motions, etc. Supporting calculations illustrating how the loads were derived are to be provided.
(3) Details and drawings of all primary structural members and offshore access gangway supporting structure (i.e., pedestal, foundation, etc.).

(4) Suitably referenced stress diagram, stress and fatigue analysis, and other supporting calculations. Where computer assisted analysis is used for the determination of scantlings, details of the software, describing input and output data, and procedures are to be included together with the basic design criteria.
(5) Material specifications.
(6) Welding details and plans indicating extent and location of non-destructive inspections of welds for gangway structure, pedestal, and foundation.
(7) Gangway pedestal and foundation drawings together with calculations indicating the maximum reactions and overturning moments.
(8) Wire rope specifications and applicable corresponding wire rope reeving diagram.
(9) Swing circle assembly drawing and details, including, as applicable:
   - Hold down bolt size with calculations, arrangement of bolts, material, grades, and pre-tensioning, together with the method used for pre-tensioning,
   - Slewing ring drawings, along with static strength calculations and details, which include material specifications of raceways and rollers or balls, hardness and heat treatment details of raceways and rollers, number and diameter of rollers or balls, raceway static capacity, specified planarity (flatness) tolerances and surface finish of bearing and supporting flanges, bearing wear tolerances, and
   - Procedure for wear down measurement of slewing ring (i.e., "rocking test").
(10) Documentation identifying proof load testing weights, locations, conditions, and procedures in accordance with Chap. 5, Sect. 5.3.
(11) Documentation of gangway risk assessment (refer to Chap. 6).
(12) The following plans, together with supporting data and particulars, are to be submitted as applicable by the asset owner:
   - Escape and access route plan,
   - Fire control plan, and
   - Emergency preparedness manual.

2.2.2.2 Offshore Access Gangway Machinery, Piping and Electric Systems

(1) Description and general details of safety devices and features, such as limit switches, auto brakes, monitoring and alarms, etc.
(2) Detailed diagrammatic plans of piping system accompanied by lists of materials giving size, wall thickness, maximum working pressure and material (including mechanical properties) of all pipes and the type, size, pressure rating and material of pumps, hoses, manifolds, valves, and fittings.
(3) Detailed diagrammatic plans of electric wiring systems including complete feeder lists, type of wire or cable, rating or setting of circuit breakers, rating of fuses and switches and interrupting capacity of circuit breakers and fuses.
(4) Documentation for computer-based systems.

(5) Details of accumulators, heat exchangers and slewing, luffing, and telescoping hydraulic cylinders indicating shell, heads, pistons, piston rods, log attachments, tie rod dimensions, and threading details, as applicable with material specifications (including mechanical properties).
(6) Detailed diagrammatic plans of central control mechanism and motion compensation mechanism including control cabinets, motion sensors, hydraulic cylinders, reduction gears and coupling bolts, and foundation arrangements, as applicable.
(7) Design justification including component strength calculations, stress analysis, material specifications, weld procedure specifications, and the extent of non-destructive examination as considered necessary are to be submitted for items (6) and (7) above.
(8) Details or manufacturer's affidavits of all prime movers such as diesel engines, motors, and generators.
(9) List of the assembled loose gear specifying the Safe Working Load (SWL) for each component.
(10) A list/booklet identifying all equipment of the offshore access gangway in hazardous areas and the particulars of the equipment; including manufacturers' names, model designations, rating (flammable gas group and temperature class), the method of protection (flammable proof, intrinsically safe, etc.), any restrictions in their use and document of certification.
(11) A declaration for absence of asbestos in the manufacture or packaging of all materials, components, equipment, machinery, piping systems and electric installations.
(12) Details of emergency and emergency recovery operational procedures, including conditions, precautions, and limitations for personnel transfer.
(13) Documentation of Failure Modes and Effect Analysis (FMEA).

2.2.2.3 Operating Manual

An operation and maintenance manual are to be submitted for review and is to include the following:

- The system information about operation conditions.
- Operating instructions for normal and degraded operating modes, and operational, deployment/ retrieval procedures.
- Step-by-step instructions for the use of the redundancy features.
- Operational limitations, such as maximum number of persons on gangway, vessel/unit motion characteristics, wind speeds, geometrical limitations, etc., and the actions to be taken to limit gangway loads.
- Safety and contingency procedures.
- Maintenance requirements.
- Periodic testing requirements.

The critical limits are to be displayed in the control cabin as well as at the gangway ends.

2.2.2.4 Testing Procedure

The following are to be submitted to Class for review:

- Load test procedures,
- Factory acceptance test procedures,
- On-board testing procedures, and
- Rocking test.

References to applicable sections of pertinent recognised standards (such as ISO 7061, UK HSE OTO 2001–069, SOLAS II-1/3–9, IMO MSC.1/Circ. 1331, as appropriate to the testing procedure) are to be identified on the plans submitted for approval and in the accompanying documentation. If the gangway supplier is employing a third party to supply the systems or components, then relevant specifications are to be reviewed by Class before the submission of the third-party documents for acceptance.

2.3   Loading, Handling and Securing

System certification is based on the understanding that responsibility rests with the asset operator or owner for control of offshore access gangway loads; offshore access gangway handling during slewing, lifting, and telescoping loads; avoidance of improper weight distributions while operating; securing of the gangway on the vessel or unit when not in use; maintenance of the offshore access gangway; and handling and stability of the vessel or unit during operation of the gangway; and the adequacy of the structure of the other vessel or unit on which the offshore access gangways lands.

2.4   Terms and Definitions

Accidental situation. A situation involving an exceptional condition of the structure, mechanical, or electrical components. For example: impact, fire, explosion, loss of intended differential pressure, failure of critical components, etc.

Allowable stress design (ASD). Allowable, or permissible, stress design is a design method in which the stresses that develop in structures or components under maximum design loads are verified against the prescribed maximum allowable stresses of the structure or component.

Base frame. The base structure to which the offshore access gangway's main bridge is attached.

Central control cabin. The main unit that controls the offshore access gangway's electrical and mechanical systems and provides the gangway operating functions, and system monitoring, etc.

Computer-based system. A system of one or more microprocessors, associated software, peripherals, and interfaces.

Design service temperature. The minimum anticipated temperature at which the gangway will be operated, as specified by the asset owner, manufacturer, or builder.

Emergency lift off accumulator package. The stored power/energy system for an emergency disconnect in case of a black out. It also can be replaced by (external) emergency power supply (e.g., from the vessel).

Heave motion compensation system. A system that is used to maintain the vertical motion of the gangway within operating limits. This includes passive or active heave motion compensation systems. A passive heave motion compensation system uses stored energy to maintain the vertical position of the gangway within the pre-set range. An active heave motion compensation system uses motion sensors and external energy to maintain the vertical position of the gangway at a predetermined location within a fixed frame of reference.

Landing structure. A component used to facilitate landing support of the telescopic bridge onto the connected vessel or unit.

Landing support structure. The support structure that attaches the landing cone support onto the connected vessel or unit.

Limit state. The point at which the structure or component has exceeded the allowable design criteria (e.g., strength, fatigue, survivability, and deflection).

Luffing cylinders. The hydraulic cylinders providing luffing movement of the gangway.

Motion compensation system. A system that used to compensate the effect of vessel motions (1 to 6 degrees of motion) on the offshore access gangway structure.

Offshore access gangway parking or stowage arrangements. The supporting structure for the gangway in its parked position on the vessel or unit deck.

Offshore access gangway system. An offshore access gangway system is a walkway that has one fixed end on the vessel or offshore unit and a telescopic end that extends to and may be temporarily attached to, another vessel or offshore unit. The system is intended to connect two vessels or offshore units primarily for personnel transfer.

Pedestal and foundations. The supporting structure above which the following are mounted: the slewing bearing assembly, the revolving upper structure, or motion compensation system.

Primary structural member or critical component. A member or component whose failure would impair the structural integrity and/or result in loss of control of the offshore access gangway. Refer to Tables 2.1 and 2.2 for examples.

Slew bearing assembly (swing circle assembly). Slew bearing (swing circle) assembly is the connecting component between the offshore access gangway's revolving upper structure and the pedestal. This component allows gangway horizontal rotation and sustains the moment, radial, and axial loads imposed by the gangway's operations.

Telescopic bridge. A telescopic frame structure used for extension of the gangway.

Table 2.1 Examples of primary structure

No	Member
1	Telescopic main bridge, telescopic bridge, and landing cone
2	Telescopic base frame (revolving frame and tube-structure), slew column
3	Eye plates, lugs, and brackets
4	Slew bearing (swing circle) assembly
5	Landing support structures
6	Offshore access gangway parking or stowage structure
7	Pins and shafts
8	Gangway supporting foundation, pedestal
9	Fasteners in the load path of all primary structural members

Table 2.2 Examples of critical machinery components

No	Component
1	Torque transmitting components of luffing, slewing and telescoping mechanism, such as drive motors, winches, drums, shafts, gears, bearings, and brakes
2	Luffing, slewing, and telescoping hydraulic cylinders
3	Motion compensation system and hydraulic cylinder

Telescopic drive. The machinery providing telescoping movement for the telescopic bridge.

Telescopic main bridge. The main frame of the gangway with one end attached to the slewing and luffing machinery. It is the pathway through which the telescopic bridge extends.

2.5 Certifications of Components

Offshore access gangway components are to be certified in accordance with Table 2.3. For applicable requirements for each component, refer to the respective chapters of this guide.

Table 2.3 Offshore access gangway components certification[a]

Component[b]		Class design review	Class unit certification	Additional notes
1	Certified safe electrical equipment			Type-tested certified by a competent, independent testing laboratory for compliance with IEC publication 60,079 or equivalent or class type approved
2	Electrical cables			Construction to be in accordance with the standards specified by Ccass
3	Sensors and communications system			Type-tested certified by a competent, independent testing laboratory for compliance with IEC publication 60,079 or equivalent or class type approved
4	Electrical motors ≥ 100 kw [c]	●	●	
5	Electrical motors < 100kw [c]	●		Test certificate furnished by the manufacturer Testing witnessed by the surveyor after installed of the offshore access gangway
6	Flexible hoses and hose end fittings	●		Design approved by class
7	Slewing, luffing, telescoping winches/gears ≥ 100 kw [c]	●	●	
8	Slewing, luffing, telescoping winches/gears < 100 kw [c]	●		Integrated gear boxes are to be design verified if located between the braking safety device and the load
9	Critical hydraulic cylinders (including piston rods) [c]	●	●	

(continued)

2.5 Certifications of Components

Table 2.3 (continued)

Component[b]		Class design review	Class unit certification	Additional notes
10	All other hydraulic cylinders (including piston rods)	●		Design review in accordance with the marine vessel classification rules
11	Internal combustion engines $\geq$ 100 kw [c]	●	●	
12	Internal combustion engines < 100 kw [c]			Manufacturer's affidavit for compliance with good commercial and marine practice testing witnessed by the surveyor after installed of the offshore access gangway
13	Loose gear			Testing as per the class guidance for lifting appliances (Lifting, and certificate furnished by the manufacturer
14	Pressure vessels and heat exchangers of 150 mm (6 in) in diameter and over and accumulators, regardless of their diameter [d]	●	●	Certification in accordance with the marine vessel classification rules
15	Pressure vessels and heat exchangers under 150 mm (6 in) in diameter			Acceptance based on manufacturer's guarantee of physical properties and suitability for the intended service, provided the installation is carried out to the satisfaction of the surveyor

(continued)

Table 2.3 (continued)

Component[b]		Class design review	Class unit certification	Additional notes
16	Slew bearing (swing circle)	●	●	
17	Wire ropes			Certificate of test furnished by the manufacture, as per Chap. 3, Sect. 3.9

Notes
[a] For materials' certification, refer to Chap. 3, Sect. 3.3
[b] For components not covered by this table, refer to the appropriate chapters and sections of this guide
[c] Applicable only for critical components, refer to Table 2.2. For non-critical components, refer to Chap. 4, Sect. 4.11
[d] Applicable only for pressure vessels and heat exchangers having design pressure, temperature and volume parameters determined by Class

Structural Requirements 3

3.1 General

This chapter outlines the structural design criteria for an offshore access gangway. While usually not required for the certification of the offshore access gangway itself, the designer should also be aware of other Rules, standards and regulations that can influence the design, such as IMO MSC.1/Circ. 1331, SOLAS II-1/3–9, UK HSE OTO 2001–069, ISO 7061:1993, ISO 7364:1983, etc. For further details, refer to the References.

3.2 Materials

All materials are to be suitable for the design environmental conditions. Materials and welding must comply with the requisite Class Rules for Materials and Welding for requirements not specifically given in the following sections. Application of international standards such as ISO 10042, EN 1090–3, EN 1999 Eurocode 9, AA Aluminium Design Manual are to be agreed upon with Class prior to the design and fabrication.

3.2.1 Steel

Material for steel structural members and components of the offshore access gangways is to comply with the requirements of the Class mandated code for lifting appliances. The primary structural members and components of offshore access gangways that are considered critically stressed are to have the following minimum thickness and effective corrosion control:

- Solid Sections: 6 mm (0.24 in) thick,
- Hollow Sections (e.g., tubular bracing): 4 mm (0.16 in) thick, and
- For less stressed members, a minimum thickness of 4 mm (0.16 in) is to be provided.

The interior of hollow sections is to be coated or proven watertight to the satisfaction of the attending surveyor.

3.2.2 Aluminium Alloys

Aluminium materials are to satisfy recognised international standards such as EN 1090–3, EN 1999 Eurocode 9, AA Aluminium Design Manual, etc., where the effects of the heat affected zones, standard commercial sections, material grades, etc. are to be suitably considered. If EN 1999 Eurocode 9 is used, then appropriately increased design factors for offshore structures are to be used as per Norsok N-001.

The fatigue of structural details in an aluminium structure is to be checked as per EN 1999 Eurocode 9 (Part 1–3), or an equivalent standard. If the code specified S–N curves do not address a specific structural detail, then the designer is to provide appropriate stress concentration factors, which are to be based on appropriate experimental results or suitable finite element analysis.

The use of aluminium alloys in offshore access gangway structures will be considered upon submission of the proposed specification for the alloy and the method of fabrication.

3.2.3 Effective Corrosion Control

3.2.3.1 Steel
Special protective coatings are to be applied to those structural members of the offshore access gangways where the thickness is less than 6 mm (0.24 in) to the satisfaction of the attending surveyor.

3.2.3.2 Aluminium
Corrosion protection of aluminium materials is to satisfy recognised international standards.

3.3 Design Loads and Conditions

3.3.1 Design Loads

The following loadings are to be considered, as applicable. When Coastal States have higher requirements that exceed the minimum requirements outlined in this reference, those requirements are to be incorporated into the design. The maximum operational length is to be used as the length of loaded gangway.

3.3.1.1 Dead Loads

The self-weight of the gangway, including landing fixture (cone), cabins, hydraulic, electrical, and mechanical items, etc. If the gangway is also intended to be utilised as a hose or temporary piping support between offshore units, the weight of hose or piping loads and hydraulic items are to be considered.

3.3.1.2 Live Loads

For a Type I gangway, the minimum live loads of 4.51 kN/m^2 (460 kgf/m^2, 94.2 lbf/ft^2) are to be applied for global design and 5 kN/m^2 (510 kg/m^2, 104.5 lbf/ft^2) are to be used for local design, refer to Fig. 3.1.

For emergency lift-off condition, a live load of 5 kN/m^2 (510 kg/m^2, 104.4 lbf/ft^2) locally distributed on the telescoping part of the gangway, refer to Fig. 3.2a or a minimum live load 15 kN (1.53 tf, 3.37 Ltf) applied on the gangway tip is considered, refer to Fig. 3.2b, whichever is greater.

For a Type II gangway, when the gangway is supported at two ends, due to a limited number of persons allowed on the gangway, the minimum design live load is to be

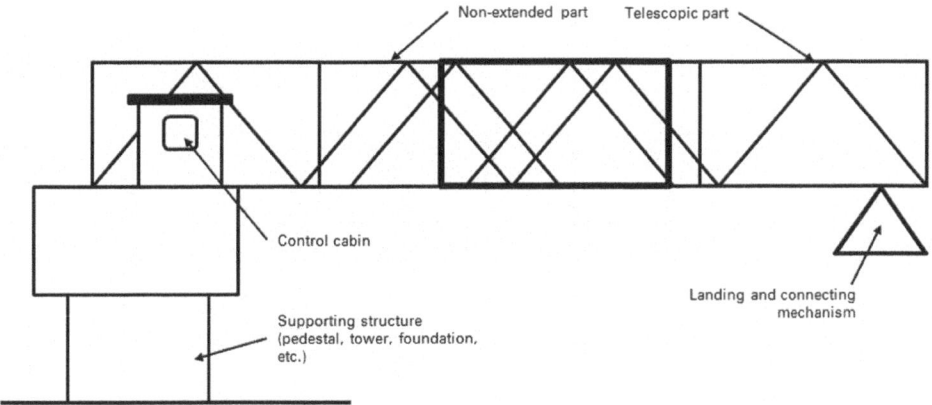

Fig. 3.1 Type I gangway: distributed live loads

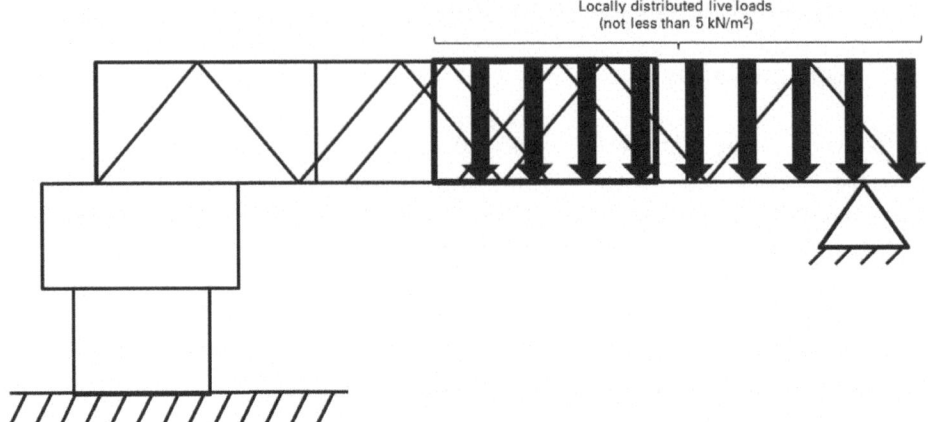

Fig. 3.2 Type I gangway: live loads for emergency lift-off condition

two times the maximum number of persons including their carry-on equipment on the gangway, refer to Fig. 3.3.

When the gangway is in the uplift or cantilever condition, the design live load is to be taken as two times the maximum number of persons including their carry-on equipment, but not less than 2.4 kN (244 kgf, 538 lbf) on the extended end, refer to Fig. 3.4.

For emergency lift-off condition, two times the maximum number of persons with one person in a stretcher or a minimum live load of 3.5 kN (357 kgf, 787 lbf) is to be applied on the gangway tip, refer to Fig. 3.5, whichever is greater.

The number of persons, weights of persons and carried items, and their location on the gangway is to be agreed upon between Class, the manufacturer, and the asset owner. The live loads on other locations of the gangway, e.g., waiting area and stairway, are to

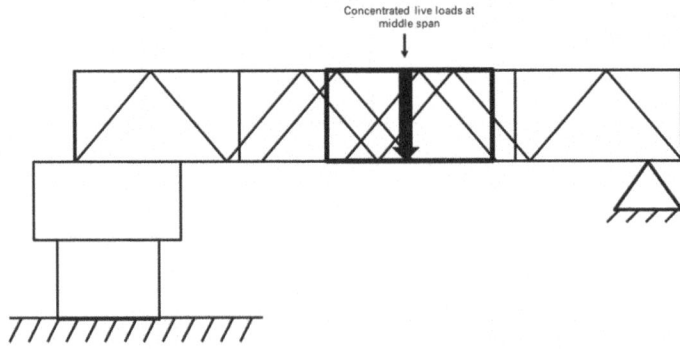

Fig. 3.3 Type II gangway: live load for gangway supported at two ends

3.3 Design Loads and Conditions

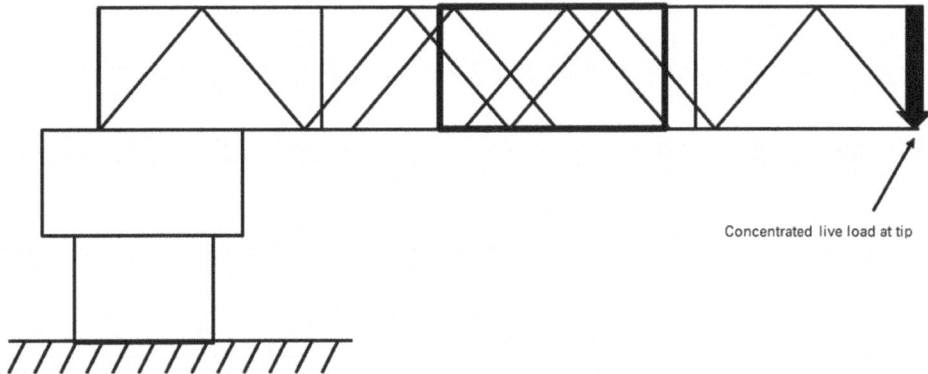

Fig. 3.4 Type II gangway: live load for uplift or cantilever condition

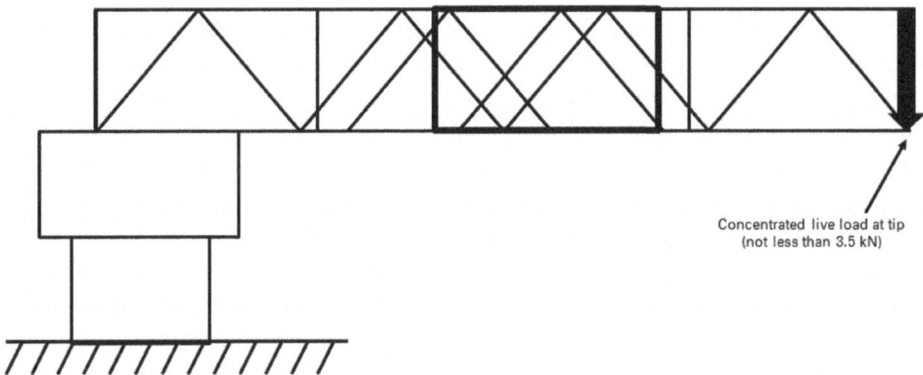

Fig. 3.5 Type II gangway: live load for emergency lift-off condition

comply with the Class Rules relating to the building and classing mobile offshore units *(MOU Rules)*.

3.3.1.3 Motion-Induced Loads

Motion-induced loads are produced by the motions of the vessel/unit on which the gangway is installed and the relative motions between vessels or units when the gangway is deployed. They are specific to the vessel/unit, the location of the gangway on the vessel/unit, and the environmental conditions where the vessel/unit operates.

The basis of certification are conditions (motions and motion-induced loads) provided by the asset owner or designer that have been established by a competent authority considering a specific vessel/unit and gangway location. The limiting environmental conditions (e.g., sea states) producing the motion effects used in the gangway's design are to be specified by the asset owner or designer. As a minimum, the selected limiting sea states are to have a return period of one year.

The asset owner or designer is usually required to specify the following:

- Magnitudes of the motion-induced vertical, transversal, and horizontal displacements and accelerations,
- Phasing of these motion-induced effects, and
- Loads produced by these motions considering each phase of the gangway's Operating, Transit, and Severe Storm Conditions (refer to Chap. 3, Sect. 3.2.3).

In the absence of specific details, the vertical, transversal, and horizontal accelerations and loads may be calculated in accordance with the Class guide for lifting appliances. The effects of motion compensation devices are to be considered, as appropriate.

3.3.1.4 Functional Loads

There may be additional global and local functional loads due to dynamic effects from lifting, lowering, slewing, telescoping, landing, angular acceleration/deceleration of rotating machinery and personnel transfer across the gangway. These functional loads are to be specified by the asset owner or designer for each operating phase. The functional loads are to be suitably combined with co-existing vessel/unit motion-induced loads. The combination loads that produce the most unfavourable load effects for the strength of the gangway in each operating phase are to be considered in design.

Monitoring is to be provided to alert the gangway operating personnel that the limits of acceptable vessel/unit motions are being approached for each of the operating phases. The margin between a design limit and a value warranting imminent action (manual or automated) to limit the loads of the gangway are to be specified by the asset owner or designer.

3.3.1.5 Wind Loads

Generally, wind loading calculations are to comply with the Class MOU Rules. The design wind velocity is based on the 1-min average gust wind velocity at the gangway location and the wind profiles are to be considered in calculating the wind load. For the normal operating condition, the design wind speed should not be less than 20 m/s (38.9 kn) for Type II gangways and 25.7 m/s (50 kn) for Type I gangways. The minimum wind velocity for all normal deployment, retrieval operations and transit conditions are not to be less than 36 m/s (70 kn).

For frame structural components, an assessment of the possibility of vortex-induced vibration due to wind on the gangway is to be included. The loads from vortex shedding and fatigue damage are to be considered for the components experiencing the vortex-induced vibration.

3.3.1.6 Ice and Snow Loads

A uniformly distributed loading of 490 N/m^2 (50 kgf/m^2, 10.5 lbf/ft^2), representing wet snow or ice, is to be considered, if applicable. Refer to the Class MOU Rules.

3.3.1.7 Impact Loads

These are loads resulting from 'green water' in the Transit and Severe Storm loading conditions described in this Chap. 3. It is considered that there will be no direct wave loading on the gangway or its supporting structure during the operating condition. Any additional loading due to direct waves on the gangway, or its supporting structure, is to be clearly specified and taken into consideration as accidental loads.

3.3.1.8 Accidental Loads

Accidental loads are loads caused by:

- Collision,
- Dropped objects,
- Fire,
- Blast, and/or
- Extreme environmental event and its induced extreme vessel/unit motions.

For a gangway that employs motion compensation system, a single failure of any component is to be treated as an event producing accidental loads. Accidental loads are to be determined on the basis of related risk assessment.

3.3.1.9 Miscellaneous Loads

Miscellaneous loads include those resulting from tie-downs or lashings used to secure the gangway in its stowed positions for Severe Storm and Transit conditions, and to provide restraints in the Operating condition. Due consideration is to be given to the redundancy of gangway lashings.

3.3.2 Design Conditions

The design is to consider the relevant Operational, Transit, Severe Storm, and Accident/Damaged conditions described below, refer to Table 3.1.

3.3.2.1 Operating Condition

The operating condition consists of three phases as follows:

(1) *Deployment phase.* This phase consists of the short-term operations required to arrange the gangway for use. The operations include movement from the stowed

Table 3.1 Design loads applied to the design conditions

Design conditions		Dead loads	Live loads	Motion-induced loads	Functional loads	Wind loads	Ice and snow loads	Impact loads	Accidental loads	Miscellaneous loads
Operating conditions	Deployment	●		●	●	Transit wind				●
	Operating static	●	●				●			●
	Operating combined	●	●	●	●	●	●			●
	Retrieval	●		●	●					
	Retrieval unexpected lift-off	●	●	●	●					
Transit conditions		●		●		Transit wind	●	●		●
Severe storm condition		●		●		Storm wind	●	●		●
Accidental conditions	Emergency lift-off	●	●	●	●	●			●	
	Damaged	●	●	●	●	●			●	●
	Impact	●	●	●	●	●		●		

Note X indicates that the load is to be applied in the loading conditions

position, slewing, luffing, extension, landing, and, as necessary, installing supplementary supports or fastenings. The loads to be considered are the dead, motion-induced, and wind loads appropriate to this phase.
(2) *Operating phase.* This phase consists of the long-term use of the gangway for personnel transfer. In this phase the gangway is considered to be either supported at both ends or operated as a cantilever. The telescopic part of the gangway may be free to move in the longitudinal direction within the fixed part. The effects from supplementary restraints are to be considered in the design (including those from tie-downs and bumpers) and, as applicable, the longitudinal compressive force from a device used to maintain such a force in the gangway.

The load cases to be considered for this phase are as follows:

- *Combined load case.* Dead, live, motion-induced, functional, and wind loads appropriate to this phase. Depending on the design of Type II gangways, additional bumper loads applied in vertical, telescopic, and transverse directions are to be considered as applicable and specified by the designer. Upon consultation with the asset owner or designer, miscellaneous loads related to the specific design may also need to be considered; these include loads arising from the static list and trim of the vessel/ unit due, and ice and snow loads.
- *Static load case.* Maximum dead and uniform live loads.
- *Retrieval phase.* This phase consists of the reverse of the deployment phase, so that at the end of this phase the gangway is configured for the Transit or Severe Storm conditions, which are described below. The loads to be considered are the dead, motion-induced, functional, and wind appropriate to this phase. If the loads for the Retrieval and Deployment phases are the same, they may be treated as one. Additionally, retrieval after an unexpected lift-off, while personnel are on the gangway, is to be evaluated. The loads to be considered are the live, dead, motion-induced, functional, and wind loads.

3.3.2.2 Transit Condition
This condition reflects the situation when the gangway is fully stowed and fastened to cradle while the vessel/unit is in transit. The loads to be considered are the dead, motion-induced, and wind loads appropriate to this phase.

3.3.2.3 Severe Storm Condition
This condition reflects the situation when the gangway is in its fully stowed and restrained condition. The loads to be considered are those resulting from the maximum storm condition that the vessel is expected to encounter during its service life. Loads due to 'green water' impact, and snow and ice are to be included as applicable based on anticipated location of operation.

3.3.2.4 Accidental Conditions

Loads resulting from the following conditions are to be considered in the design:

(1) *Emergency lift-off condition.* The gangway's extended end lifts off in its full-length condition with maximum dead, motion-induced, and wind loads appropriate to the Operating phase; plus, the Functional load reflecting the lift condition. The gangway's centrifugal/radial effects from luffing and slewing are to be considered and to be agreed by Class. A live load applied on the uplift or cantilever free end is also to be considered in combination with the other mentioned categories of loads.
(2) *Damage to mechanical components.* The loads resulting from the failure of any single mechanical component (e.g., hydraulic cylinder, wire rope pulley, etc.) that provides support to the gangway when the gangway is arranged for its Operating phase. For example, in the case of a failure of a hydraulic system hose, a gangway may lose the support of luffing cylinders. Also, the failure of any motion compensation device is to be considered as a damaged condition.
(3) *Impact load case.* To account for random impacts on the primary structural members of the gangway, a minimum load of 5 kN (0.51tf, 0.52Ltf) is to be applied in any load bearing structure, anywhere along the span of the gangway when it is arranged for its operating phase.

3.4 Strength Assessment

3.4.1 General

The offshore access gangway structure covered in the scope of this reference guide is as specified in Chap. 2, Sect. 2.1. All the pertinent structural components are to be analysed considering the design conditions specified in Sect. 3.2. The various categories of structure, as indicated in Table 3.2, are to be assessed using appropriate linear elastic methods to determine the adequacy of the structure.

The design acceptance criteria are concerned with four limit states as follows:

- Accidental Limit States (ALS) to better verify the survival of the structure when subjected to anticipated accidental and damaged conditions,
- Ultimate Limit States (ULS) to resist yielding, buckling, and ultimate strength,
- Fatigue Limit State (FLS) to resist fracture from cyclic load effects, and
- Serviceability Limit State (SLS) to address the structural deflections of the gangway.

The adequacy of the structure to resist the four mentioned limit states is to be demonstrated by appropriate structural analysis supplemented, as necessary, by testing. Structural

3.4 Strength Assessment

Table 3.2 Structural strength assessment

		Yielding check	Buckling check	Ultimate strength check	Fatigue check
Local structures	Plating	●	●	●a	–
	Stiffeners	●	●	●a	–
Primary structural members		●	●	●	●b
Hull interface structures		●	●c	●c	●d

Notes "X" indicates that the strength assessment is to be carried out
[a] The ultimate strength check is included as part of the buckling check
[b] The fatigue check of primary structural members is the fatigue check of connection details of these members
[c] The buckling and ultimate strength check is to follow the Class rules for buckling and ultimate strength assessment for offshore structures (Class buckling guidance)
[d] The fatigue check of the steel structure is to follow the Class guide for fatigue assessment for offshore structures (Class fatigue guidance)

analyses and checks against the limit state criteria are to be performed using applicable and proven techniques and software.

3.4.2 Allowable Stress Assessment Criteria for ULS and ALS

The ASD criteria given herein are specifically for gangway components made of structural steels. The application of ASD assessment criteria requires the determination of representative allowable stresses for individual components. Allowable stresses are not to be exceeded for the type of component and loading condition being considered.

3.4.2.1 Allowable Stresses

Computed tensile, bending and shear stress components and, as applicable, combinations of such stresses, for primary structural members, stresses are not to exceed the allowable stress, F, as obtained from the following equation:

$$F = F_y \bullet S_c$$

where?

F_y = specified minimum yield strength of material. For design purposes, for steels with yield strength not exceeding 355 N/mm^2 (36 kgf/mm^2, 51 ksi) F_y is to be limited to no more than 72% of the minimum ultimate strength of the steel.

S_c = allowable stress coefficient as specified in Table 3.3. For accidental conditions, an increase of up to 33% in the allowable stresses may be used.

Table 3.3 Allowable stress coefficients, S_c

Type of stress	Allowable stress coefficients, S_c
Tension	
Non-pin connected members (gross area)	0.60
Pin connected members (net area)[d]	0.45
Shear	
On the cross-sectional area effective in resisting shear	0.40
Bending (tension and compression on extreme fibres)	
Solid round and square bars	0.75
Members with compact sections[c]	0.66
Members with non-compact sections[c]	0.60
Bearing stress	
On contact area of surfaces and projected area of pins in holes	0.90
Combined stress	
Von Mises stress (static loads)	0.67
Von Mises stress (combined loads)	0.75

Notes
[a] Members subjected to combined stresses are to be proportioned to satisfy the requirements of Chap. 4, Sects. 4.2.1 and 4.2.2
[b] For additional guidance, refer to the AISC Specifications for the Design, Fabrication and Erection of Structural Steel for Buildings, June 1, 1989. Other recognised references can be used provided the ultimate strength of the component is limited to linear elastic behaviour as is typically done in the mentioned references
[c] For classification of sections as compact or non-compact, refer to the Class buckling guide
[d] For non-redundant critical pin connected members, consideration may be given for higher safety factors to account for the increased risk associated with personnel use. Refer to Chap. 4, Sect. 4.6

For plated structures, von Mises stress using finite element analysis for all load conditions is not to exceed $F_y \bullet S_c$, where S_c is the allowable stress coefficient.

$S_c = 0.67$ for operating condition–static loads.
$= 0.75$ for operating condition–combined loads
$= 0.75$ for transit condition and severe storm condition
$= 0.90$ for accident/damaged condition

The requirements for appropriate mesh size in the finite element analysis refer to the Class guidance notes on safe hull finite element analysis of hull structures.

3.4.2.2 Buckling and Ultimate Strength

The representative buckling and ultimate strength of the individual members and plates comprising the gangway's principal structure is to be based on the criteria provided in the Class buckling guide. For allowable stress design the allowable stress criteria is typically

3.4 Strength Assessment

given as a fraction of the material's yield stress, the component's buckling stress, or as an Interaction Ratio that accounts for combined stresses (e.g., bending with axial tension or compression).

- For the Operating Conditions described in Sect. 3.2.1, the maximum values for these fractions are as specified in the Class buckling guide and the AISC 1989 Specification. In the Class buckling guide, the fractions applicable to 'Normal Operations' of an offshore installation and 'Static Loading' of a MODU apply. For AISC, the fraction values stated in the Specification apply (i.e., without a one-third increase in stress). For 'Combined Loading' conditions, an increase of up to 10% in the allowable stress may be used. A typical value of the fraction, say for axial tension on the gross cross-section of an individual member, is 0.6.
- For the Transit Condition and the Severe Storm Condition described in Sect. 3.2.2, the values of the fractions must correspond in the Class buckling guide to the fractions applicable to 'Severe Storm' for an Offshore Installation and 'Combined Loadings' for a Mobile Offshore Unit; in the AISC 1989 Specification, 'General Provisions'. For Severe Storm Condition, an increase of up to 10% in the allowable stress may be used.
- For the Accidental Conditions described in Sect. 3.2.3, the values of the fraction can be increased to 0.8 of the component's characteristic strength.

When plates are used as principal structure, the total stress resulting from orthogonal membrane and shear stress components can be determined using the von Mises criteria. The applicable allowable stresses for each Design Condition are as given above. If the plate is subjected to normal pressure loads refer to the Class buckling guidance. Also refer to the Class buckling guidance for stiffened plates and shells.

3.4.3 Fatigue Assessment Criteria for FLS

Fatigue assessments of the structural details in the main structure of the gangway are to be performed. The characteristic strength of steel structural details as located in the Class fatigue guidance, which gives S–N curves pertinent to the expected fatigue performance. The criteria in the Class fatigue guidance can be supplemented by other recognised standards when dealing with issues such as bolted connections, cast and forged components, and structural details made of materials other than steel.

The fatigue of structural details in an aluminium structure is to be checked as per EN 1999 Eurocode 9 (Part 1–3), or an equivalent standard. If the nominal S–N curves do not address a specific structural detail then the design is to be based on appropriate experimental results, or suitable Finite Element Analysis, to determine stress concentration factors to be used with the 'hot-spot' S–N curve.

Table 3.4 Fatigue design factors

Importance	Inspectable and field repairable	
	Yes	No
Non-critical	2	5
Critical	3	10

Note "Critical" implies that failure of these structural items would result in the rapid loss of structural integrity and produce an event of unacceptable consequence

Fatigue damage is the result of varying stress in the structural detail. The loading conditions that are to be considered in the fatigue assessment are those specified in Sect. 3.2. In consideration of the intended service of the gangway and the vessel on which it is located, the asset designer or asset owner is to submit an anticipated loading 'history' for the gangway. Conditions representing the anticipated Accidental Conditions may be omitted from the fatigue assessment provided a thorough inspection emphasising the critical details will be performed if a major accident or structural failure occurs and that the incident and inspection results are promptly reported to Class.

The cumulative fatigue damage and the corresponding fatigue life can be estimated by the Palmgren–Miner linear damage rule. The minimum required fatigue life of a structural detail is related to the design life of gangway structures by a Fatigue Design Factor that depend on the inspectability, repairability, redundancy, the ability to predict failure damage, as well as the consequence of failure of the structural detail. The minimum required Fatigue Design Factors to be obtained in the design of a structural detail are given below in Table 3.4.

3.4.4 Serviceability Limit State (SLS)

The relative deflection of the gangway, δ_{max}, in the operating condition, is not to exceed the following criteria:

$\delta_{max} \leq \frac{\ell}{100}$ Gangway designed with a cantilever free end.

$\leq \frac{\ell}{200}$ Gangway designed with both supported ends.

Where:

δ_{max} = maximum relative vertical or lateral deflection.
= $max(|\delta_{B1}|, |\delta_{B2}|)$
= δ_{B1} and δ_{B2} are shown in Sect. 3.3, Fig. 3.6
ℓ = design length of the gangway.

3.4 Strength Assessment

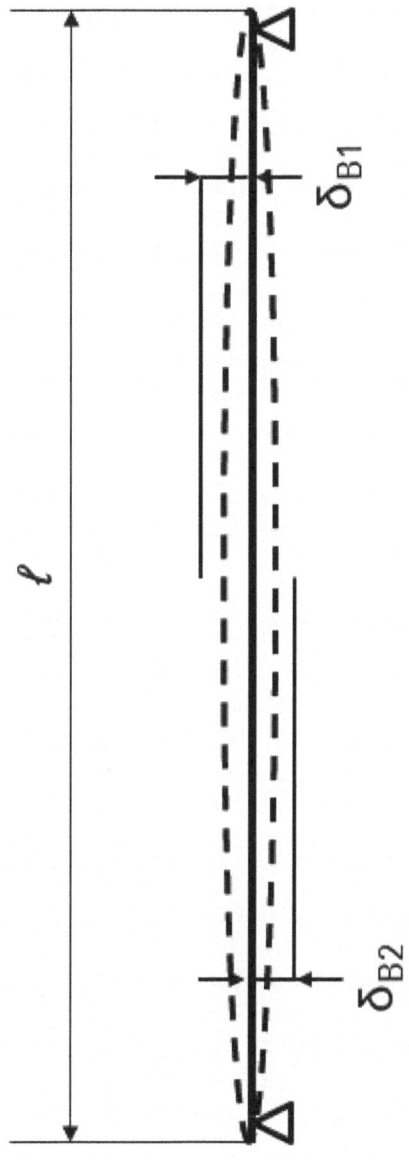

Fig. 3.6 Relative deflection

3.5 Landing Mechanism

The landing device (e.g., landing cone) and its supporting structure at the telescoping end of the gangway are to be designed for the landing and connection loads for the operating and accidental conditions specified in Sect. 3.2. The landing mechanism design is also to include the appropriate maximum interface loads for landing platform design, the adequacy of the emergency release system, the adequacy of the lashing system, the adequacy of the shock absorption system and the end attachments such as a landing platform. The minimum landing loads are to be the resultant loads in telescoping, slewing, and luffing directions at the gangway tip due to the gangway dead loads, live loads, wind loads, functional loads, and vessel motion induced loads.

The gangway lashing release system is to be suitably robust to allow for the emergency disconnect scenario in the event the accelerations exceed the allowable limiting values or if the motion compensation (e.g., shock absorber) stroke is exceeded. Allowable stresses for the landing device and its support in the gangway structure are given in Sect. 3.4. The appropriate rigging safety factors in the lashings and release mechanism are to follow the appropriate Class rules.

3.6 Pedestals, Foundations, and Supporting Structures

The pedestal, foundation, and supporting structures in the primary load path (e.g., pedestal shell, bearing support, internal structures, pump, and engine room, etc.) are to be designed for all anticipated design loading conditions in accordance with Sect. 3.2, where the horizontal (longitudinal and transverse) and vertical loads, including the applicable dynamic amplification factors, are to be multiplied by the pedestal factor 1.5. The strength criteria of the pedestal, foundation, and associated supporting structures are to be in accordance with Chap. 4.

The live loads on walkways, work areas, and laydown areas attached to the pedestal are to be taken as per Sect. 3.1.2.

For non-redundant critical items (e.g., pins, hydraulic cylinders, etc.) that connect gangways to the lifting mechanism or pedestals, a higher factor of safety is to be applied to account for the increased risk associated with personnel use. This is to be determined by appropriate risk assessment or, in the absence of specific recommendations from the risk analysis, the non-redundant critical items are to be designed with a minimum factor of safety of 7.5 (with 2 pins sharing the design load) for a gangway in operation.

Detail drawings and strength assessments of pedestals, foundations and supporting structures are to be installed are to be submitted and approved prior to certification.

3.7 Slewing Mechanism

Slewing ring (swing circle) assembly including slewing ring, bearing, bolting, and gears in the slewing mechanism are to satisfy the requirements of the Class guidance for lifting appliances.

3.8 Telescoping Mechanism

The design of telescoping mechanism and the design check of its components, such as rollers, bushings, drag chain, etc., are to be in accordance with applicable international standards and practices, such as EN 1993–6 Part5, EN 13,001-3-1 Annex C.4.

Drag chains can be treated as structural components subject to normal structural practice together with some nominal standards for chain-type systems. These items may not affect structural integrity but are used to protect/separate the hoses/cables during the telescoping motion of the gangway.

3.9 Wire Rope

Wire rope used in the offshore access gangway is to comply with the requirements of the Class guidance for lifting appliances.

3.10 Loose Gears

Loose gear used in the offshore access gangway is to comply with the requirements Class guidance for lifting appliances.

Machinery and Systems

4.1 General

The mechanical, piping, electrical, and safety systems and components of offshore access gangways that are used for lifting, slewing, luffing, telescoping, and landing (including lashing and securing) systems are subject to design review for compliance with the requirements of this chapter, and as applicable, the Class mandated MOU Rules and marine vessel classification rules.

4.2 Materials

Materials for machinery systems and components are to comply with Chap. 3, Sect. 3.2.

4.3 Electrical Systems

Electrical systems are to be designed, constructed, installed, and tested to the requirements outlined in this reference guide and, as applicable, the appropriate sections of the marine vessel classification rules and/or the Class MOU Rules.

4.4 Piping Systems

In general, piping systems are to be designed, constructed, installed, and tested to the requirements contained in the appropriate sections of the marine vessel classification rules and/or the Class MOU Rules.

4.4.1 Hydraulic Systems

Hydraulic systems are to be designed, constructed, installed, and tested to the requirements, as specified in the appropriate sections of the marine vessel classification rules and/or the Class MOU Rules.

4.4.1.1 Hydraulic Oil Tanks
Hydraulic oil tanks are to meet the requirements of the appropriate sections of the marine vessel classification rules and/or the Class MOU Rules.

4.4.1.2 Hydraulic Cylinders
Hydraulic cylinders that are used for luffing, slewing, and telescoping and all other cylinders that are considered as critical, in accordance with Chap. 2 , Table 2.2, are to be designed, constructed, and tested to the requirements of the Class guidance for lifting appliances.

Where cylinders are used for luffing, slewing, and telescoping, each motion is to be provided with one of following:

(1) One cylinder with double seals at the piston head end rod, and
(2) Two independent cylinders, where each cylinder is to be independently capable of holding the personnel using the gangway.

All other cylinders are to be designed to the requirements of the appropriate sections of the marine vessel classification rules and/or the Class MOU Rules.

4.5 Pressure Vessels

Pressure vessels are to be designed, constructed, installed, and tested to the requirements, as applicable, of the appropriate sections of the marine vessel classification rules and/or the Class MOU Rules.

4.6 Rotating Machines

Internal combustion engines, electrical motors, generators, and other rotating machines whose failure would not result in loss of control of the offshore access gangway, are to be designed, constructed, and equipped in accordance with good commercial and marine practice and are to meet the design requirements of the offshore access gangway for items such as operating temperature, duty cycle, and angle of inclination, as specified in the asset designer's specification. Such equipment need not to be inspected at the plant of the manufacturer, but will be accepted based on manufacturer's affidavit, verification of the nameplate data, and satisfactory performance testing witnessed by the surveyor after installation.

Internal combustion engines, electrical motors, generators, and other rotating machines that are considered as critical components as per Chap. 2, Sect. 2.4 must comply with Chap. 2, Table 2.3. Redundancy is to be provided for the critical components that need power and control for active motion compensation systems.

For the design requirements of rotating machines, winches and gearboxes refer to the applicable Class guide for lifting appliances.

4.7 Computer-Based Control Systems

Where fitted, a computer-based control system of the offshore access gangway is to comply with the requirements of the appropriate sections of the marine vessel classification rules and/or the Class MOU Rules.

4.8 Motion Compensation Systems

The motion compensation system installed on the offshore access gangway may include either passive or active motion compensation. The requirements for the motion compensation system are to satisfy the requirements of the applicable Class guide for lifting appliances or equivalent.

For an active motion compensation system, a fully redundant power management system is to be provided, and all electrical equipment that is part of the active motion compensation system is to be built with redundancy so that a single failure will not completely disable the functions of the gangway.

Documentation demonstrating redundancy of these systems is to be submitted for review.

Where the gangway consists of an active motion compensation system or a motion sensing system that gauges or controls the relative motions between the gangway and connected vessel, a suitable monitoring system (or a closed-circuit television) is to be provided to monitor the gangway operation. When the gangway loses contact with the connected vessel, an alarm is to be triggered and the gangway is to be returned to the safe position to avoid injury to personnel. A detailed Failure Modes and Effects Analysis (FMEA) or other equivalent risk analysis is to be required to demonstrate the active motion compensation system or the motion sensing system is fault tolerant. The analysis is to be submitted for review.

4.9 Low Temperature Operation

For an offshore access gangway with a design service temperature below −10 °C (+14°F), the manufacturers of the machinery systems are to demonstrate by way of testing or analysis that these systems will operate satisfactorily at the design service temperature. Critical machinery components are to comply with the applicable Class guide for lifting appliances.

4.10 Hazardous Locations

Electrical equipment, including all electrical power, control, and safety devices, and wiring on the gangway installed in hazardous locations (where a flammable atmosphere may exist) are to be suitable for operation in such areas, and are to comply with the appropriate sections of the marine vessel classification rules and/or the Class MOU Rules. Where essential for operational purposes, internal combustion engines and mechanical equipment may be installed in hazardous areas subject to special consideration.

In general, exhaust outlets are to discharge outside of all hazardous areas, air intakes are to be not less than 3 m (10 ft) from hazardous areas and any parts of equipment whose surface temperature may exceed 200 °C (392°F) are to be effectively insulated, cooled, or protected by other means.

4.11 Blocking Mechanism

A blocking mechanism is to be installed to block the hydraulic, electric, or pneumatic circuit, as applicable, if the gangway luffing or slewing angles exceed their limits, and to prevent the offshore access gangway from "running out".

When the gangway is in its operating mode, the following functions are to be maintained:

(1) All brakes are to be fail-safe to maintain positive control of the load at all times. A risk assessment for brakes acting in normal operating and in case of emergency, emergency stop, and power failure, etc. situations is to be submitted for Class review,
(2) Where fitted, all monitoring systems, all overload protection systems, and automatic/manual protection systems are not to be overridden and locked out, and
(3) Where fitted, motion compensation systems including active heave compensation systems and passive heave compensation systems are not to be overridden and locked out.

The system requirements and the electrical and hydraulic schematics of the blocking mechanism are to be in accordance with the appropriate sections of the marine vessel classification rules and/or the Class MOU Rules.

4.12 Electronics and Communications

The control cabin communication systems, control systems, and electronics, including sensors integrating various individual systems (such as in lifting mechanism, slewing mechanism, and telescoping mechanism), are to be designed and fabricated in accordance with the appropriate sections of the marine vessel classification rules and/or the Class MOU Rules.

4.13 Emergency Recovery

The gangway is to be provided with a means to recover the personnel from gangway bridges in the event of a single failure in the power or control system. When personnel are unable to disembark using a walkway or ladder, a secondary power supply system and an independent control system for main functions (e.g., slewing, luffing, telescoping, etc.) may be used for this purpose.

Components (such as pipes, flexible hoses, and electric cables) that are used only for transfer of power or signals from the power unit to the actuators (motors, cylinders, etc.), need not be taken into consideration in the single failure of the power and control system.

The manual activation switches or handles for the secondary system are to be of a "hold to run type" and clearly and permanently marked for their purpose and are to be in a location with a clear view of the gangway's operations.

The emergency recovery function is not to be affected by the loss of main power.

An instruction document giving detailed instructions is to be provided at the operator's station for all procedures.

4.14 Safety Systems and Arrangements

4.14.1 General

The arrangements for safety systems are to comply with appropriate sections of the marine vessel classification rules and/or the Class MOU Rules and/or IMO MSC.1/Circ.1331 as detailed below.

4.14.2 Monitoring Systems

Suitable monitoring systems are to be provided to constantly monitor, display, record, and save the followings in real time at 1 Hz to the system database:

- Wind speed,
- Vessel accelerations and spacing between connected vessels/units, and
- Gangway movements including luffing and slewing angle, telescoping distance, etc.

4.14.3 Alarm System

Audible and visual alarms are to be provided on the gangway and another manned location in the following situations:

- The gangway operational limits are being approached,
- The vessel/unit motion limits are being approached at the warning level margins as indicated in Chap. 3, Sect. 3.1.4,
- If fitted, the gangway's overload protection system indicates a warning or hazard, and
- Loss of power or emergency recovery.

The alarm system may be integrated with the gangway monitoring systems.

4.14.4 Handrails and Grids

Handrails are to be provided on both sides of the gangway. The protection grids or handrails are to be provided for relative movement between gangway sections due to sliding components.

Handrails are to comply with the following:

(1) Height of at least 1 m (39 in),
(2) Provided with at least three courses, i.e.:
   - The opening below the lowest is not to exceed 230 mm (9 in),
   - Other courses are not to be more than 380 mm (15 in) apart, and
(3) Stanchions are not to be more than 1.5 m (60 in) apart.

Additionally, the handrails are to be able to withstand an impact load of 750 N/m (76.5 kgf/m, 51.4 lbf/ft) at the upper guide level without permanent deformation.

4.14.5 Slip Resistant Surface

All gangways are to have slip-resistant surfaces and treads in compliance with IMO MSC.1/Circ.1331.

4.14.6 Landing Area

For gangways supported at both ends, the connected vessel or unit shall provide an appropriate landing platform, side shock absorption system, or equivalent arrangements to prevent unacceptable longitudinal, transverse, and vertical movement from gangway landing devices.

If a gangway designed to operate as a cantilever, the gangway is to provide a mechanism to keep the tip end in a standing position in all three dimensions with a tolerance of less than 100 mm (3.9 in) for the distance.

4.14.7 Break-Away System

The gangway landing cone or connection end to the landing platform or supporting structure shall have an appropriate break-away system to allow the gangway to be easily disconnected from the landing platform. The function of the break-away system shall not be affected by the loss of electrical power.

4.14.8 Fire Protection

The gangway is to be made of non-combustible material generally. The firefighting systems and fire protection requirements, and integrity of the hoistway enclosure are to be in accordance with the appropriate sections of the marine vessel classification rules and/or the Class MOU Rules.

4.14.9 Control Cabin Protection

The fire protection requirements of the control cabin are to be in accordance with the appropriate sections of the marine vessel classification rules and/or the Class MOU Rules. If the cabin is the operation station hosting the operator, the cabin is to be designed to protect from weather and other environmental exposure and provide the operator appropriate operation view and living conditions.

4.15 Visual Aids

4.15.1 Markings

Restrictions on the safe operation and loading, including the range of permitted design angles of inclination, design load, etc., is to be prominently displayed at each end of the gangway in compliance with IMO MSC.1/Circ.1331, Sect. 3.5. Additionally, for a Type II gangway, the maximum number of persons allowed to use the gangway at the same time is to be included.

The steps and edges of gangway's walkways are to be clearly delineated. Signs or markings are to be provided at the end of gangway's walkway. And at the transition to the telescoping part of the gangway, the markings are to be clearly visible by both day and night, clear and unambiguous. Yellow or white road paint can be used for this purpose.

4.15.2 Lighting System

Lighting systems, including normal and emergency/warning lights, are to comply with the appropriate sections of the marine vessel classification rules and/or the Class MOU Rules. Adequate lighting is to be provided at the means of embarkation and disembarkation, the immediate deck area of embarkation and disembarkation, the length of gangway, and the controls of the arrangement.

Testing and Surveys 5

5.1 General

This chapter outlines the survey requirements during construction, installation, and after construction for offshore access gangways placed on board a vessel or unit. During construction and before being placed in service, new gangways are to be subjected to acceptance tests and inspections at the manufacturing facility and on the vessel/unit to verify compliance with the requirements of this reference guide and Class approved plans.

All acceptance testing and surveys are to be witnessed and accepted by a Class approved surveyor. Testing as required by the Flag State for the vessel/unit may also be witnessed and monitored by the Class approved surveyor.

5.2 Surveys During Construction

5.2.1 General

All gangways are to be surveyed during construction to the extent necessary for the surveyor to determine that the details, material, welding, and workmanship are acceptable to Class and are in accordance with the approved drawings.

Non-destructive testing is to be carried out in accordance with Class approved plans to the satisfaction of the attending surveyor.

5.2.1.1 Quality Control System
The manufacturer shall establish and maintain a quality control system to verify that all Class requirements, including design approval, materials, verification, fabrication workmanship, and non-destructive testing are complete.

The quality control system should provide sufficient details of manufacturing and inspection to verify that the manufacturer's inspections are performed at appropriate stages of fabrication. In the event of non-compliance, fabrication should be delayed for rectification.

The quality control system should fully document welding procedures and qualification of welding personnel. The quality control system should also detail the procedures and qualifications of non-destructive testing personnel to be employed in all stages of fabrication and manufacture.

5.2.2 Functionality Testing

To demonstrate gangway functionality, the final testing of the gangway is to include load testing as per Class approved procedures along with the following, as appropriate:

(1) Functional test of motion compensation system,
(2) Functional test of telescoping system,
(3) Functional test of luffing system,
(4) Functional test of slewing system,
(5) Emergency lift system,
(6) Testing of electrical and communication systems,
(7) Testing of safety systems,
(8) Testing of monitoring and alarms,
(9) Testing of control systems,
(10) Testing of landing system,
(11) Testing of gangway disconnection arrangement, and
(12) Fault simulation test for the redundancy arrangement.

5.3 Load Testing

Load testing conditions are to be identified for each gangway component based on the most severe loading conditions, and a load testing procedure that identifies the test loads is to be submitted for Class review.

5.3.1 Load Testing Procedures

The load testing is to be conducted as follows:

5.3 Load Testing

(1) The specified minimum test time is to be twice the time to reach the final target set-down position from the resting cradle, or vice versa, whichever is longer. The test time is to be a minimum of five minutes. For example, if the estimated set-down time is three minutes, then the testing duration is to be six minutes.

In the case of a gangway with continuous axial compression, the load testing is to be performed while the gangway is connected at both ends with the continuous force maintained, in accordance with Class approved procedures.

(2) The test load is to be determined for various design conditions, in accordance with Class approved procedures, including the live loads on the walkway, the effect of the hanging loads, landing cones, etc.
(3) Apply the test loads as per Class approved procedures at the outer end of inner fixed part of the gangway, lift the gangway to the horizontal position, and then extend the gangway. Move the gangway to the extents allowable for slewing (swinging) and luffing (raising and lowering). After the specified test time, retract the gangway and remove the load.
(4) Apply the test loads as per Class approved procedures, at the outer end of outer telescopic part of the gangway, lift the gangway to the horizontal position, and then extend the gangway. Move the gangway up and down to the extents allowable for slewing (swinging) and luffing (raising and lowering). After the specified test time, retract the gangway and remove the load.
(5) The Class surveyor will verify the following:
- System leakage during testing.
- Visual inspection of the gangway after the test for deformation, excessive wear or fractures especially in way of critical elements (e.g., non-redundant elements such as pins) as identified in the testing procedure. *Non-destructive* testing of components for fractures is to be in accordance with the Class approved test procedures. Additional locations may be selected by the attending Class surveyor for further inspection or testing.
- Maximum vertical and lateral deflections of the gangway are to be recorded for each test scenario, with and without the test loads; and deflections are to be checked against allowable values noted in the Class approved test procedures.
- Maximum angles of operation.
- The accelerations of the gangway during the lifting/lowering/slewing operations are to be recorded and compared against the design values. If the values exceed the design assumptions, then suitable corrective actions are to be taken.

5.3.2 Test Loads

5.3.2.1 Test Loads for Gangway Bridge

(1) *Type I Gangway*. When the gangway is extended to its maximum operational length and supported on both ends, a test load that equals to 1.25 times design live loads [4.51 kN/m^2 (460 kgf/m^2, 94.2 lbf/ft^2)] is to be applied along the gangway.

The maximum relative deflection of the gangway should not exceed L/200 and no permanent deformation is to be noted.

(2) *Type II Gangway*. For a gangway designed to be supported at both ends, when it is extended to its maximum operational length, a test load that equals to the maximum design live loads (not less than 2.4kN (244kgf, 538lbf)) multiplied by a dynamic amplification factor is to be applied at the middle of the gangway. The dynamic amplification factor is not to be taken less than 1.25.

For a gangway designed to operate as a cantilever, when it is extended to its maximum operational length, a test load equal to the maximum design live loads times the dynamic amplification factor, but not less than 3 kN (305 kgf, 672.5 lbf), is to be applied at the tip of the gangway. The dynamic amplification factor is not to be taken less than 1.25. For the gangway designed for carrying more than two persons, the test load distribution along the gangway is to be agreed by Class.

The maximum relative deflection of the gangway should not exceed L/200 for the gangway supported on both ends and L/100 for the gangway operated as a cantilever, and no permanent deformation of the gangway is to be noted.

5.3.2.2 Test Loads for Gangway in Uplift Position

To simulate the gangway lift-off or loss of support at one end and achieve the maximum overturning moment at the slewing bearing, the gangway is to be uplifted in cantilever position and operated to its maximum operational length.

(1) Case 1. No personnel carried in the uplift position:

The test load is to be applied at the extended end of the gangway:
  Test load $= SW \times DAF - 1.0 \times L_{cog}/L$
  Where:
  SW = self-weight of gangway.
  DAF = dynamic amplification factor, it is not to be taken less than 1.25.
  L_{cog} = distance between the gangway support centre and the centre of gravity of gangway at maximum extension.

L = maximum operational length.

(2) Case 2. Personnel carried in the uplift position:

For type I gangway:
The test load is to be applied at the extended end of the gangway:
Test load = $SW \times DAF - 1.0 + LL \times L \times W \times L_{cog}/L$
Where:
SW = self-weight of gangway.
LL = live loads with 5 kN/m² (510 kg/m², 104.5 lbf/ft²) distributed evenly along the length of the gangway.
DAF = dynamic amplification factor, it is not to be taken less than 1.25.
L_{cog} = distance between the gangway support centre and the centre of gravity of gangway at maximum extension.
L = maximum operational length.
W = width of the gangway bridge.

For type II gangway:
The test load is to be applied at the extended end of the gangway:
Test load = $SW \times DAF - 1.0 \times L_{cog}/L + DAF \times LL$
Where:
SW = self-weight of gangway.
LL = design live load but not less than 1.2 kN (122 kgf, 269 lbf).
DAF = dynamic amplification factor, it is not to be taken less than 1.25.
L_{cog} = distance between the gangway support centre and the centre of gravity of gangway at maximum extension.
L = maximum operational length.

Alternative test loads and load application locations may be accepted by agreed with Class.

5.4 Surveys During Installation

Prior to the gangway being placed into service, a Class surveyor is to attend the vessel or unit to verify the initial installation is in accordance with Class approved drawings and to examine the following:

(1) Structural attachment of the gangway and associated supporting structures to the vessel/unit,
(2) For gangways fitted with slewing rings:

(a) Prior to mounting of the gangway, the surveyor is to witness flatness checks and surface finish requirements to verify compliance with the manufacturer's specifications for the following:
- Gangway attachment area for slewing ring,
- Slewing ring, and
- Mounting flange on pedestal,

(b) Shimming or surface levelling compounds are not to be used to attain the required level of flatness of the mounting surfaces,

(c) During installation, slew ring bolts are to be pretensioned by controlled means or alternative means (e.g., prototype testing, electronic measuring, etc.). Pretensioning, by bolt torque or by hydraulic tensioning device, is to be in accordance with the bearing manufacturer's instructions, which are to be submitted for review. Elongation of the bolts is to be measured to verify pre-tensioning. At least 10 percent of the bolts, randomly selected, are to be measured to the satisfaction of the attending surveyor,

(d) After the gangway has been mounted, a "Rocking Test" is to be carried out in accordance with the bearing manufacturer's instructions and the results are to be included in the Register,

(3) Testing of the piping system in accordance with this guidance, and satisfactory installation, including protection of hoses, if any,

(4) Electrical wiring, including wiring in fixed cable trays, electrical equipment in hazardous areas, and terminations,

(5) Function testing of the gangway, including all limits and operational modes, and

(6) Load testing of the gangway in accordance with Chap. 5, Sect. 5.3, and test conditions and results should be included in the Register of the offshore access gangway; refer to Chap. 6.

Function test of safety protective devices for the power source and prime movers.

5.5 Surveys After Construction

In addition to the regular inspections before and after use of the gangway, the following periodic inspections are to be carried out.

5.5.1 Annual Survey

After undergoing the initial installation survey and examination required by Sect. 5.4, the offshore access gangway is required to undergo an Annual Survey at intervals of not more

5.5 Surveys After Construction

than 12 months. The following are to be examined and placed in satisfactory condition as found necessary and reported upon:

(1) Functional testing and non-destructive testing (NDT) of the critical elements (i.e., non-redundant elements) as noted in the Class approved plans,
(2) Visual inspection of the gangway structure for deformation, excessive wear, corrosion, damage, and fractures; NDT for factures as appropriate,
(3) Visual examination of foundations for deformation, excessive wear, corrosion, damage, and fractures,
(4) Visual external examination and operational test of gangway machinery including prime mover, clutches, brakes, slewing and luffing machinery, hydraulic system, and safety valve settings,
(5) Visual examination and operational condition of the gangway's telescoping system including sliding surfaces,
(6) Visual inspection of wire rope including end attachments,
(7) Visual aids inspection, as appropriate,
(8) The slewing ring assembly, where applicable, is to be examined for slack bolts, damaged bearings and deformation as required by the manufacturer,
(9) On-board function testing of the gangway, including all limits and operational modes, and
(10) All safety devices and fire protection systems are to be tested and personnel emergency recovery performed in accordance with the submitted manufacturer's procedures.

5.5.2 Retesting Survey

At intervals of five years, in addition to the requirements of the Annual Survey in Sect. 5.1 above, the gangway is to undergo testing and examination as follows:

(1) Prior to load testing, for gangways fitted with a slewing ring, the surveyor is to witness a Rocking Test in accordance with the bearing manufacturer's recommendations, and a grease sample is to be analysed. Twenty percent of the slewing ring bolts are to be removed and non-destructively tested. Bolts chosen for examination are to be taken from the most highly loaded area of the slewing ring and their position is to be noted for future surveys. If any bolts are found with defects, additional bolts are to be removed to confirm suitability for continued use. If the results of the Rocking Test and grease samples indicate bearing wear in excess of the manufacturer's recommendation, the bearing is to be opened for internal examination. Alternative methods of testing of the slewing ring and bolts may be specially considered.

(2) Retesting Surveys are to include the load testing required for Installation Survey in Sect. 5.3.

Upon completion of the load testing, the slewing ring including bolting arrangements and foundation are to be examined for slack bolts, damaged bearings, and deformed or fractured weldments. As deemed necessary by the surveyor, further analysis of slewing ring grease samples for metal particles and NDT examination of the slewing ring for fractures or damage may be required.

(3) Retesting Surveys are to include functional testing followed by visual examination, as well as NDT, of the critical elements as appropriate. Upon completion of functional tests, the critical welds of gangway's pedestals or kingposts are subject to the following non-destructive testing to the satisfaction of the attending surveyor:
- Volumetric NDT of all critical butt welds in the gangway's pedestals or kingposts, including any transition pieces between the pedestal and the slewing ring. If both sides are accessible and 100% volumetric NDT has been previously completed and recorded in the gangway's records, 100% surface NDT on both sides may be conducted instead,
- 100% surface NDT on both sides of critical fillet welds in the pedestal or kingpost and transition pieces,

(4) A close-up examination of all structure, luffing structural connections, multiple sheave blocks, spreaders, hydraulic cylinders, and all other load bearing parts is to be carried out to confirm their condition. Suitable safe means of access are to be provided to facilitate the close-up examination. Any load-carrying parts that display indications of damage or deformation are to be further examined as deemed necessary by the attending surveyor.

5.5.3 Repairs and Alterations

5.5.3.1 Telescopic Structure and Permanent Fittings

When repairs or renewals, including welding and/or replacement of major structural components, are required to be made to the load bearing structures or permanent fittings of the offshore access gangways, the repairs are to be carried out to the satisfaction of the surveyor. Any welding is to be one by an approved procedure. Tests and examination of the gangway are to be carried out in accordance with Sect. 5.2 of this chapter. All load tests are to be conducted unless the manufacturer identifies the load test required to test the repair or modification. If all load tests are conducted, the asset owner is to consider conducting a retesting survey.

The repairs or renewals are to be noted in the gangway Register, and repair reports are to be attached to the certificate as an Appendix.

Examples of load bearing structures requiring retest are:

(1) Telescopic main bridge, telescopic bridge, and landing structure,
(2) Telescopic base frame, slew column,
(3) Pedestal and foundation,
(4) Swing circle (slew bearing) assembly, and
(5) Pins and shafts.

5.5.3.2 Modification of Gangways

If the offshore access gangway is repaired with different materials, rebuilt with profiles of a different cross-section, or structural components changed, the gangway design may need to be resubmitted to Class for approval. A new test and examination may be required to be conducted in accordance with Sects. 5.2, 5.3, and 5.4 of this chapter. If the modifications are found satisfactory, the Surveyor may issue a new certificate in accordance with Chapter 7, Sect. 7.2.

5.5.4 Reinstallation Survey

For gangways that have been reinstalled to a different vessel/unit, installation and retesting surveys are to be conducted, including a review of the maintenance records.

5.6 Slewing Ring Surveys

The slewing ring surveys are to follow the requirements of the Class guidance for lifting appliances.

5.7 Inspection of Wire Rope

Wire ropes are to be inspected at each annual and retesting survey in accordance with the Class guidance for lifting appliances. The gangway owner is to examine the wire rope at frequent intervals between surveys.

5.8 Monthly Inspection by Vessel's Personnel

A monthly inspection of the offshore access gangway is to be made by members of the vessels or unit's personnel as designed by the Master. A record is to be kept of the findings of the inspections, along with any repairs and renewals resulting from these inspections. This record is to be in or kept with the offshore access gangway Register.

Risk Assessment and Register for Offshore Access Gangways

6.1 General

Considering the complexity of the gangway systems (especially for active motion compensated offshore access gangways), in each phase of operation, system risk assessment through Failure Mode Effect and Criticality Analysis (FMECA) is to be performed to verify appropriate consideration has been given to critical component failure and that sufficient redundancy is available where components are not failsafe.

Where components are of a new/novel type, or are being used in a completely novel manner, component FMECA may be required to demonstrate component suitability or as input to the system risk assessment.

Class will typically provide detailed guidance and reference on matters pertaining to risk assessment applications for the marine and offshore industries, which will provide the guidelines for defining the concept of risks, describing the methods available to assess the risk associated with offshore units and setting up and conducting successful risk studies.

The gangway supporting utility functions and primary escape routes may require a holistic risk assessment plan which involve performing a HAZID/HAZOP study for the purposes of generating a hazard register, and further studies as necessary in the detailed design phase (e.g., fire and explosion analyses; emergency system survivability analysis, smoke and gas ingress analysis; Escape, Evacuation, and Rescue (EER) study; quantitative risk assessments (QRA)).

Risk assessment can be performed by a third party. The risk analyses are primarily addressed to those items affecting the safety of an installation, facility, or operation, but the methods discussed can also be applied to other types of risk. The risk analysis findings are to be incorporated into the relevant manuals and test procedures.

The FMECA or risk studies performed on a case-by-case basis, as applicable, by considering the relevant aspects from offshore units that are to be connected by the gangway and any secondary functional aspects (e.g., support of utility transfer lines). Any specific FMECA or risk study is to demonstrate that no single failure will result in unacceptable consequences, such as gangway unavailability, potential to result in personnel injury, environmental impact, or equipment damage.

6.2 Register for Offshore Access Gangways

A Register for the offshore access gangway is to be available onboard for endorsement by the surveyor at the time of periodic and damage surveys; refer to Chap. 5. The following items are to be included in the Register: arrangement diagram of the assembled gangway, loose gear location and marking list, operation manual, particulars and location of special materials, welding procedures, and a record of periodical surveys. Additionally, copies of certificates covering original and replacement loose gear, original tests to the gangways and repairs or additions to the gangways are to be attached to the Register.

6.3 Certificates and Forms

The following certificates and forms are usually provided by the builder, manufacturer, testing authority, or the firm undertaking annealing (when required). Copies, as required and appropriate in each case, are to be made available for inclusion in the Register.

- Certificate of test and examination of chains, rings, loose gears, shackles, swivels and pulley blocks,
- Certificate of examination and test of wire rope before being taken into use,
- Manufacturer's bolt and torque standards for slewing ring bearing,
- Approved corresponding wire rope reeving diagrams, and
- Manufacturer's procedures for proof-testing of hydraulic cylinders including overriding of limiting devices (where required) to achieve full proof load.

The following forms and reports are provided and issued by the Surveyors (as applicable) upon completion of prescribed tests and surveys. Copies are to be included in the Register.

- Register of offshore access gangway,
- Certificates of Test and examination of offshore access,
- Gangway and their accessory gear/cylinders: before being taken into use,
- Retesting surveys and tests associated with repairs,
- Certificate of annual thorough examination of cylinders and for annual inspection of offshore access gangway, and

- Reports covering the construction of the offshore access gangway and any tests carried out at the manufacturer's plant during construction.

6.4 Asset Owner's Overhaul and Inspection Record

A record is to be kept onboard the vessel or unit to show particulars of all overhauls, inspections, repairs, and replacements carried out by the offshore access gangway owner or operator. This record is to be made available to the surveyor at all times and, in addition to the above requirements, is to have specific sections that include the following:

- A log of the "Rocking Testing" results outlined in Chap. 5, Sect. 6.3, showing the manufacturer's tolerances and remaining slew bearing clearances calculated from the testing results,
- A record of the slew bolts inspected, as required by Chap. 5, showing the location of bolts and a copy of bolt manufacturing record or certificate, if the bolts have been renewed, and
- A copy of the NDT record of all critical weld inspections after load testing, as required by Chap. 5.

6.5 Repairs and Alterations

Certificates covering tests performed after repairs and alterations are to be inserted in the Register.

6.6 Additions of New Gear and Wire Rope

Replacement wire rope and loose gear is to be supplied with manufacturer's certificate conforming to tests in accordance with the Class guidance for lifting appliances. The wire rope and loose gear certificates are to be inserted in the Register and each article and certificate is to be identified as to location in the gangway assembly. Certificates covering discarded loose gear are to be removed from the Register.

Appendix

Samples of Register of Offshore Access Gangways

AMERICAN BUREAU OF SHIPPING

REGISTER OF OFFSHORE ACCESS GANGWAY

NUMBER OF REGISTER BOOK _____

DATE OF ISSUE _____

PORT OF ISSUE _____

NAME OF VESSEL _____

PORT OF REGISTRY _____

IMO/OFFICIAL NUMBER _____

OWNER _____

ADDRESS _____

SAMPLE ONLY NOT TO BE USED

CHG-1 GRC

REGISTER OF OFFSHORE ACCESS GANGWAY

INSTRUCTIONS

1. This Register of Offshore Access Gangway is issued in connection with the ABS *Guide for Certification of Offshore Access Gangways* and is to be kept available for inspection of proper authority and endorsement by the Surveyor at the time of inspections.

2. The Register is divided into three parts for the purpose of recording the following information:

    PART I - The Surveyors are to fill in the required information with respect to the original load tests and examination of the vessel's offshore access gangway in accordance with Section 5, "Load Testing", "Functionality Testing", "Survey during Installation", "Slewing Ring Surveys", and "Inspection of Wire Rope", and with respect to Annual and Retesting Surveys of the Offshore Access Gangways on the vessel in accordance with 5/9.

    PART II - A record shall be kept in this section of the monthly inspection of the offshore access gangway made by the vessel's personnel as required by 5/15.

    PART III - In this part, there shall be inserted the following certificates of tests, examinations, and inspections if relevant:

    a. Reports of Test and Examination of Offshore Access Gangway at manufacture.

    b. Certificate of Test and Examination of Wire Rope. Form CHG-5, Refer to CHG-5 Form, A-1 of the *ABS Guide for Certification of Lifting Appliances*.

    c. Certificate of Test and Examination of chains, rings, loose gears, shackles, swivels and pulley blocks. Form CHG-4, Refer to CHG-4 Form, A-1 of the *ABS Guide for Certification of Lifting Appliances*.

    d. Certificate of Initial Test and Retesting, or Tests Associated with Repairs. Form CHG-3-GRC

    e. Certificate of Annual Examinations and Special Inspections. Form CHG-7-GRC

    On the reverse side of the above mentioned certificates will be found the particulars of tests pertaining to each.

Appendix   59

NOTES ON SPECIAL MATERIALS FOR THE PRINCIPAL STRUCTURAL PARTS

Telescopic Main Bridge, Telescopic Bridge or Landing Supporting Structures: -

Base Frame, Slew Column, Swing Circle Assembly, Pedestal, Pins and Shafts, Fasteners : -

SAMPLE ONLY
NOT TO BE USED

Other Structural Parts and Components: -

CHG-1 GRC

Vessel's Name _____

PART I
INITIAL TEST AND SUBSEQUENT ANNUAL AND RETEST
INSPECTION CERTIFICATES

THIS IS TO CERTIFY that the gangway listed below has been surveyed and found in satisfactory condition unless otherwise noted under Remarks. (If all of the gangways are surveyed at the same time, it will suffice to so indicate below; however, if this is not the case, each article or unit inspected should be listed.)

DESCRIPTION AND LOCATION OF GANGWAY	DATE OF SURVEY	NO. OF CERTIFICATE	SIGNATURE OF SURVEYOR	REMARKS

SAMPLE ONLY
NOT TO BE USED

CHG-1 GRC

Appendix

Vessel's Name _____

DESCRIPTION AND LOCATION OF GANGWAY	DATE OF SURVEY	NO. OF CERTIFICATE	SIGNATURE OF SURVEYOR	REMARKS

CHG-1 GRC

Vessel's Name _____

PART II
RECORD OF MONTHLY INSPECTION BY VESSEL'S PERSONNEL

DATE INSPECTED	REPAIRS AND RENEWALS REQUIRED RESULTING FROM INSPECTION	SIGNATURE OF VESSEL'S PERSONNEL

SAMPLE ONLY
NOT TO BE USED

CHG-1 GRC

Appendix

Vessel's Name _____

PART II

RECORD OF MONTHLY INSPECTION BY VESSEL'S PERSONNEL

DATE INSPECTED	REPAIRS AND RENEWALS REQUIRED RESULTING FROM INSPECTION	SIGNATURE OF VESSEL'S PERSONNEL

SAMPLE ONLY
NOT TO BE USED

CHG-1 GRC

Vessel's Name _____

PART II
RECORD OF MONTHLY INSPECTION BY VESSEL'S PERSONNEL

DATE INSPECTED	REPAIRS AND RENEWALS REQUIRED RESULTING FROM INSPECTION	SIGNATURE OF VESSEL'S PERSONNEL

*SAMPLE ONLY
NOT TO BE USED*

CHG-1 GRC

Appendix

Vessel's Name _____

PART III

CERTIFICATES of tests, examinations, and inspections are to be inserted behind this sheet.

**SAMPLE ONLY
NOT TO BE USED**

Test Certificate No. _____

CERTIFICATE OF TEST AND EXAMINATION OF OFFSHORE ACCESS GANGWAYS AND THEIR ACCESSORY CYLINDERS AND TESTS ASSOCIATED WITH REPAIRS AND RETESTING SURVEY

Name of ship on which offshore access gangway is fitted _____ Class Number _____

(1) Description and Location of Offshore Access Gangway	(2) Operation Angles or loading conditions at which the gangway was operated	(3) Test load applied	(4) Safe working load (S.W.L.)

5. Issuance for completion of Retest Survey: YES _____   NO _____

6. Issuance in association with repairs only: YES _____   NO _____

   (b) Wash box fitted: YES _____   NO _____   S.W.L. assumes wash box empty.

7. For commencement of Retest Survey of Offshore Access Gangway see Report No. _____ dated _____

<u>REMARKS</u>

This Certificate valid until: _____

8. Name and address of association witnessing the test and making the examination: **American Bureau of Shipping**
   Houston, Texas, U.S.A.

   Port of Survey _____

9. Position of signatory in association: **Surveyor to American Bureau of Shipping**.

   I certify that on the _____ day of January _____, the above cargo gear was tested by a competent person in a manner set forth on the reverse side of this certificate; that a careful examination of the said machinery and cylinder by a competent person after the test showed that it had withstood the test load without damage or deformation; and that the safe working load of said machinery and cylinder is as shown in Column 4.

   (Date) _____   _____ - Surveyor

CHG-3 GRC   In substantial agreement with I.L.O. Form No. 2

Appendix

INSTRUCTIONS

After installation or major repair and when the offshore access gangway is placed in service it shall be initially tested to a load equal to 125% of the working load of the assembled gangway, for subsequent Retest Surveys and for minor repairs it shall be tested to a load equal to 110% of the working load of the assembled gear. A general, careful examination of all accessible parts of the assembled gangway is to be carried out after the load test. Where damaged or deformed condition is noted, parts are to be further examined to determine the condition of the affected parts.

NOTE: The expression "ton" means a ton of 2240 lbs unless stated otherwise. Load is to be recorded in pounds per running foot of conveyor an also in total tons.

For the purpose of this certificate a competent person is defined as a Surveyor of a Classification Society or other recognized certificating agency.

For additional ABS requirements see Section 5 of the ABS *Guide for Certification of Offshore Access Gangways*."

SAMPLE ONLY
NOT TO BE USED

NOTE: This Certificate evidences compliance with one or more of the Rules, guides, standards or other criteria of ABS and is issued solely for the use of ABS, its committees, its clients or other authorized entities. This Certificate is a representation only that the structure, item of material, equipment, machinery or any other item covered by this Certificate has met one or more of the Rules, guides, standards or other criteria of ABS. The validity, applicability and interpretation of this Certificate is governed by the Rules and standards of ABS who shall remain the sole judge hereof. Nothing contained in this Certificate or in any Report issued in contemplation of this Certificate shall be deemed to relieve any designer, builder, owner, manufacturer, seller, supplier, repairer, operator or other entity of any warranty express or implied.

CHG-7 GRC

Certificate No. _____

CERTIFICATE OF ANNUAL THOROUGH EXAMINATION OF OFFSHORE ACCESS GANGWAY AND ITS CYLINDERS

Name of ship on which offshore access gangway is fitted

_____ Class Number _____

Description and Location of Gangway	Number and date of Certificate of last test and examination (Form CHG 3 G.L.)

REMARKS

Name and address of association witnessing the test and making the examination: **American Bureau of Shipping**
Houston, Texas, U.S.A.

Port of Survey _____

Position of signatory in association: **Surveyor to American Bureau of Shipping**.

I certify that the above offshore access gangway was thoroughly examined by a competent person and that no defects affecting its safe working condition were found other than those indicated and corrected as noted under remarks.

(Date) _____   _____
                                                                                        - Surveyor

NOTE: For the purpose of this certificate a competent person is defined as a Surveyor of a Classification Society or other recognized certificating agency.

CHG-7 GRC

Appendix

INSTRUCTIONS

The following parts are to be visually examined in place at each Annual Inspection. Dismantling of the bearings may be required where damaged or deformed condition is noted.

1. Primary or secondary load bearing structure conditions (cracks, distortions, corrosion). NDT of the critical elements may be applied as appropriate.
2. Offshore access gangway support structure.
3. Excessive clearance in sheave-bearings and eye-bolt connections.
4. Wire rope, tables, cable connections including end attachments, wear, broken wires and corrosion inspections.
5. Operation condition of slewing and luffing system (slewing/luffing bearing condition, lubrication, bolt condition and pretension, etc.).
6. Slewing and luffing equipment including safety devices and limit switches.
7. Operation condition of telescoping system, sliding surface condition.
8. Functional operation of the gangway system and motion compensation system if applicable.
9. Valves, cocks, pipes, strainers, and cylinders.
10. Leakages in hydraulic system and correct safety valve adjustment.
11. Safety systems and alarms, monitoring systems.
12. Operation condition of electrical and communication systems.
13. Safety marking and safety devices.
14. Gangway storage and parking structures, gangway disconnection arrangement.
15. Fire extinguishing system, etc.

NOTE: This Certificate evidences compliance with one or more of the Rules, guides, standards or other criteria of ABS and is issued solely for the use of ABS, its committees, its clients or other authorized entities. This Certificate is a representation only that the structure, item of material, equipment, machinery or any other item covered by this Certificate has met one or more of the Rules, guides, standards or other criteria of ABS. The validity, applicability and interpretation of this Certificate is governed by the Rules and standards of ABS who shall remain the sole judge hereof. Nothing contained in this Certificate or in any Report issued in contemplation of this Certificate shall be deemed to relieve any designer, builder, owner, manufacturer, seller, supplier, repairer, operator or other entity of any warranty express or implied.

CHG-7 GRC

References

1. BS EN 50018:2000, Electrical Apparatus for Potentially explosive Atmospheres–Flameproof Enclosure 'd'.
2. CFR 1918, Safety and Health Regulations for Longshoring.
3. EN ISO 10042:2005, Welding–Arc-welded Joints in Aluminium and its Alloys–Quality Levels for Imperfections.
4. EN 1090-3:2008, Execution of Steel Structures and Aluminium Structures–Part 3: technical Requirements for Aluminium Structures.
5. EN 1990:2002: Eurocode, Basis of Structural Design.
6. EN 1993:2006: Eurocode 3, Design of Steel Structures.
7. EN 13001-3-1: Cranes General Design.
8. EN 1993-6: Eurocode 3, Design of Steel Structures–Part 6: Crane Supporting Structures.
9. EN 1999:2007: Eurocode 9, Design of Aluminium Structures.
10. IEC 60092, Electrical Installations in Ships.
11. IEC 60228, Conductors of Insulated Cables.
12. IMO MSC.1/Circ. 1331, Guidelines for Construction, Installation, Maintenance and Inspection Survey of Means of Embarkation and Disembarkation.
13. ISO 7364:1983, Shipbuilding and Marine Structures–Deck machinery–Accommodation Ladder Winches.
14. ISO 7061:1993, Shipbuilding–Aluminium Shore Gangways for Seagoing Vessels.
15. ISO 15540:1999, Ships and Marine Technology–Fire Resistance of Hose Assemblies–Test Methods.
16. ISO 10042:2003, Welding–Arc-welded Joints in Aluminium and its Weldable Alloys–Quality levels for Imperfections.
17. ISO 19900:2013, Petroleum and natural Gas Industries–General Requirements for Offshore Structures.
18. ISO 19901-6, Petroleum and Natural Gas Industries—Specific requirements for offshore structures Part 6: Marine Operations.
19. NFPA 15, Standard for Water Spray Fixed Systems for Fire Protection.
20. Norsok N-001, Integrity of Offshore Structures, Edition 7, June 2010.
21. Norsok S-001, Technical Safety.
22. Norsok Z-015, Temporary Equipment.
23. SOLAS II-1/3-9, "Means of Embarkation on and Disembarkation from Ships."
24. UK HSE OTO 2001-069, Decks, Stairways and Their Associated Handrails.

SPRINGER NATURE

GPSR Compliance

The European Union's (EU) General Product Safety Regulation (GPSR) is a set of rules that requires consumer products to be safe and our obligations to ensure this.

If you have any concerns about our products, you can contact us on ProductSafety@springernature.com

In case Publisher is established outside the EU, the EU authorized representative is:

Springer Nature Customer Service Center GmbH
Europaplatz 3
69115 Heidelberg, Germany

The manufacturer's authorised representative in the EU is Springer Nature Customer Service Centre GmbH, Europaplatz 3, 69115 Heidelberg, Germany. If you have any concerns regarding our products, please contact ProductSafety@springernature.com

Printed and bound by CPI Group (UK) Ltd, Croydon, CR0 4YY

26/03/2026

02078953-0018